FOOD AND
AGRARIAN ORDERS
IN THE
WORLD-ECONOMY

FOOD AND AGRARIAN ORDERS IN THE WORLD-ECONOMY

EDITED BY
Philip McMichael

PRAEGER

Westport, Connecticut
London

The Library of Congress has cataloged the hardcover edition
as follows:

Food and agrarian orders in the world-economy / edited by Philip
    McMichael.
        p.   cm. — (Contributions in economics and economic history,
    ISSN 0084–9235 ; no. 160) (Studies in the political economy of the
    world-system)
        Includes bibliographical references and index.
        ISBN 0–313–29399–6 (alk. paper)
        1. Produce trade—Congresses. 2. Agriculture—Economic aspects—
    Congresses. 3. Food supply—Congresses. 4. Land tenure—
    Congresses. 5. International economic relations—Congresses.
    I. McMichael, Philip. II. Series. III. Series: Studies in the
    political economy of the world-system.
    HD9000.5.F548   1995
    338.1—dc20   94–17977

British Library Cataloguing in Publication Data is available.

A hardcover edition of *Food and Agrarian Orders in the World-Economy*
is available from Greenwood Press, an imprint of Greenwood Publishing
Group, Inc. (ISBN: 0–313–29399–6).

Library of Congress Catalog Card Number: 94–17977
ISBN: 0–275–94966–4

First published in 1995

Praeger Publishers, 88 Post Road West, Westport, CT 06881
An imprint of Greenwood Publishing Group, Inc.

Printed in the United States of America

The paper used in this book complies with the
Permanent Paper Standard issued by the National
Information Standards Organization (Z39.48–1984).

10 9 8 7 6 5 4 3 2 1

Contents

Illustrations

Introduction: Agrarian and Food Relations in the World-Economy

Philip McMichael

The chapters in this book stem from the seventeenth annual Political Economy of the World-System (PEWS) conference, held at Cornell University, Ithaca, New York, in April 1993. The conference is held under the auspices of the American Sociological Association. While the theme is always global, each year the topic (and venue) is different. In 1993 the focus was on the global dynamics of food and agricultural systems, a focus that is past due, and expresses a growing understanding of food and agrarian orders in the world-economy.

The delay in focusing on food/agriculture has an institutional explanation insofar as academic developmentalist studies embody an "urban bias." Theories of social change and modernity routinely have assigned a residual status to agriculture and rural life (Buttel and McMichael 1988), and this has contributed to the minority status of agricultural studies in the academy. World-system analysis has not been immune to this particular institutional bias, despite the powerful critique offered by Wallerstein of nineteenth-century social theory and its legacy of "developmentalism" (see, e.g. Wallerstein 1991). Developmentalism has been associated with industrialization on a country-by-country basis. The current proliferation of global manufacturing systems, however, scrambles *national* developmentalism by discounting (national) place and privileging global markets in money, labor, and goods. The resulting perception of a "developmentalist illusion" (Arrighi 1990) derives from historical insights such as Wallerstein's that "development" is a (national) organizing myth that has enslaved and obscured the vision of modern social science (Taylor 1987). This postmodern critique clearly presaged current concerns with social and environmental sustainability on a world scale.

The critique is located in Wallerstein's historical analysis. In the first of several volumes tracing the history of *The Modern World-System* from the sixteenth century, Wallerstein relates the rise of the absolute monarchies in England and France to what he terms the ''peripheralization'' of Eastern Europe and Hispanic America (1974). He argues that the Atlantic powers, emerging as centers of world commerce and manufacturing, with intensive systems of contractual agricultural labor, were complemented, and provisioned, by coerced cash-crop systems: the so-called second serfdom of grain producers on Eastern European estates, slavery on sugar production, and the *encomienda* system, which was used to control labor in Peruvian mines or Guatemalan indigo plantations. That is, the different forms of agricultural labor across the world-economy signify two central dynamics: first, an unequal division of world labor exists, and, second, the forms of agricultural labor are correlated with the relative strength of states (understood not just in politico-military terms, but also in terms of state capacity to coordinate complementary economies—whether mercantile, imperial, or national). And this has a critical corollary: The most powerful states have historically gained the capacity to set the rules, not only of international commerce, but also of discourse, including the discourse of industrial development, as the fetish of modern national civilizations. For Wallerstein, developmentalism—the master narrative of modern social thought—stems from the social organization of the world-economy. It has accordingly shaped theories of social change, relegating food/agriculture to the margins.

It is not, however, simply a question of industrial fetishism. A more fundamental issue is the general rationalization of humanity associated with the ''endless accumulation'' of capital (Wallerstein 1984; Sayer 1991). Central to this is the increasing subordination of the material world and its self-conceptions to exchange value, as all things assume a price and modern culture submits to the calculus of value. Market rationality, however, has its own limits because its universalist claims are contested—whether because the sinews of social life (labor, land, and money) cannot be fully commodified (Polanyi 1957), or because value relations have always depended on nonvalue relations (such as patriarchal arrangements, or nonwage labor relations), or because non-Western entities (states or ethnonational or subnational movements) do not embrace the imperial implications of market rationality (see, e.g., Cheru 1989; Watkins 1991). In these arenas of contested meanings lie perhaps the renewed focus on agricultural and environmental sustainability, the preservation of food cultures and food security, and, therefore, the growing interest in understanding the global dimensions of food and agricultural systems.

ORGANIZATION OF CONTENTS

These chapters reflect these interests, ranging from analyses of power and meaning in the construction and reconstruction of social diets, through situating agriculture within long-term movements of the world capitalist economy, to spatial and regulatory shifts in the organization of contemporary agro-food production systems. The table of contents of this volume follows these lines of inquiry.

The first part, "Food in World-Historical Perspective," includes three chapters that examine the process of large-scale dietary construction from symbolic, political, and technological angles. The first two chapters arose from the conference organizer's attempt to establish a dialogue concerning the multiple dimensions of power in food production and consumption—both Sidney Mintz and Harriet Friedmann were asked to address special conference sessions on the topics of their contribution. Sidney Mintz discusses the power relations that configure food preferences. He links large-scale power relations associated with the supply of specific foods and the cognitive and symbolic aspects of consumption, suggesting a mutual conditioning of structural changes and consumers' reordering of their categories of meaning. Harriet Friedmann addresses the related question of why the relations of food production and consumption assume such contemporary significance, anchoring it in the Polanyian cycle between societal self-protection and (current) market self-regulation. Superimposed on this cycle are structural trends associated with transnational corporate forms of capital accumulation, the political decline of farm sectors, and the transformation of work and consumption relations—all of which proffer new possibilities for exiting the cycle via new, localized forms of democratic regulation. The third chapter in this section, by Susan Thompson and Tadlock Cowan, reevaluates the role of "durable foods" in global food relations, through a case study of canned seafood's strategic role in provisioning military forces and industrial proletarians during the rise of the capitalist world-economy.

The second part, "Agriculture in World-Historical Perspective," juxtaposes Ravi Palat's controversial reconsideration of the historical implications of wet-rice cultivation, with Resat Kasaba and Faruk Tabak's long-term, world-systemic perspective on the sources of the collapse of the modern agrarian order. Each chapter deploys agriculture to reframe questions about world-historical social change. Palat argues that the specificities of Asian wet-rice cultivation and its reproductive networks precluded the development of agrarian capitalism, thereby challenging the Eurocentric views that capitalism was as indigenous to Asia as it was to Europe. Kasaba and Tabak examine three overlapping cycles of the world-system in the last 200-odd years—a cyclical movement of agricultural prices, the Columbian exchange between "Old" and "New Worlds" of wheat and maize, and the rise and fall of the British and American hegemonies—arguing that these, rather than simply the institutions of U.S. hegemony, stand behind the current reconfiguration of the world division of labor.

The third part, "Contemporary Agro-Food Complexes," consists of case studies of the spatial and institutional organization of four (now) universal foodstuffs: the pervasive additive carrageenan, wine, coffee, and shrimp. Lanfranco Blanchetti–Revelli's chapter examines the relocation of seaweed production as the outcome of a corporate reconstruction of the inputs to the carrageenan industry. The year-round cultivation of seaweed in the Philippines displaced former natural seaweed beds in Eastern Canada, illustrating the contradictory dynamics of global sourcing and its organization by transnational corporate processors. Walter Goldfrank and Roberto and Miguel Korzeniewicz analyze the differential success of Argentina and Chile in replacing import-substitution industrial policies with exports of such nontraditional products as fresh fruits and wine, as a strategy of negotiating zonal structures in the global wine commodity complex. John Talbot's discussion of the tropical commodity complex, centered in coffee, extends Friedmann's three postwar food complexes of wheat, meat, and durable foods. He insists on the importance of a fourth complex: the tropical commodity complex. Not only does this complex represent the remaining colonial division of labor in the postcolonial world, but also its producer states deploy cartels and alliances with local capital to constrain transnational corporate control of world markets. Finally, Mike Skladany and Craig Harris consider the role of transnational firms in reorganizing the exploding global shrimp industry along scientific lines (the "blue revolution"), emphasizing the regulatory problems associated with the occupation of (formerly common) swamplands and wetlands.

The final part, "Recomposition of Global and Regional Agro-Food Systems," examines some of the key contemporary mechanisms of restructuring. Fran Ufkes focuses on the role of the Singapore state in industrial upgrading in the 1980s. Part of this strategy involves relocating pork production for the Singapore market to agro-export platforms in surrounding Southeast Asia, and retaining well-regulated agrotechnology parks in a strategy of repositioning Singapore within the (globalized) regional economy. Jane Collins follows with an examination of the restructuring of global labor forces for agribusiness, critiquing the presumption that women's labor in fruit and vegetable production is flexible and politically quiescent. Ramon Grosfoguel situates Caribbean depeasantization within the confluence of decolonization and postwar transformation of the international division of labor. This has involved the undermining of agrarian structures and the promotion of nonagricultural development, reinforced, at the household level, by remittances from labor migrants in the United States. The concluding chapter, by Robert Schaeffer, recovers the global view, detailing the implications of current free trade agreements, including GATT and NAFTA, for agro-environmental sustainability. Further, the movement toward supranational regulatory mechanisms not only privileges transnational corporations, but also disaggregates the interstate system thereby seriously reducing poorer states' access to multilateral and bilateral concessional trade and aid.

CONCEPTUAL THREADS

Alternatively, this volume is also framed by three conceptual threads, with distinct analytical foci: the world-system, food regimes, and global commodity chains. In many respects these foci parallel Fernand Braudel's three social times: the *longue durée*, political/institutional conjunctures, and the event (Braudel 1972). Their field of vision normally involves distinct explanatory purposes, even though ideally we would envisage these social times (and forces) as mutually conditioning, as Braudel would have it. Certainly these foci are intellectual relatives, but nonetheless in practice they seek their own level. The exemplar of world-system analysis is the chapter by Kasaba and Tabak, which traces the processes involved in the construction and reconstruction of world agrarian orders. These are the cycles mentioned above (prices, growth/contraction in productive and geographical terms, and hegemony), cycles that constitute and express world-system dynamics. The overriding emphasis, on the contribution of these dynamics to cereal production concentration in the core zones of the world-economy in the post–World War II era, follows the aforementioned Wallersteinian specification of a world division of labor as the framework of modern social action, and hence of analysis.

Food regime analysis focuses on the state/capital nexus, examining the relations of food production and consumption that condition the restructuring of the interstate system and industrial capitalist social relations (Friedmann and McMichael 1989). Thus, in the late nineteenth century, the first food regime formed through the relocation of capitalist agriculture to the New World, as the new site of temperate food production for industrializing Europe. That is, increasingly capitalized settler agriculture cheapened the provisioning of metropolitan proletarians with grain and meat exports (Friedmann 1978). At the same time it anchored, outside of Europe, a new form of nationally organized state and economy (McMichael 1984). Following the interwar crisis of the twentieth century, U.S.-inspired world-economic reconstruction via decolonization established the national model of accumulation as a social *ideal*, based on agro-industrial integration and regulated through the Bretton Woods institutional complex. Food companies internationalized such integration, outgrowing subsidized metropolitan farm sectors associated with postwar Fordism. The latter incorporated consumption relations into accumulation, transforming food from a low-cost input of the wage-relation to an input in an industrial food complex servicing a high-wage consumer economy. The condition of this was the peculiar status of agriculture as a protected sector, removed from GATT consideration, and yet organized to overproduce—first in the United States and then in Europe as a consequence of the Marshall Plan–guided reconstruction of national farm sectors (Cleaver 1977; Block 1977). The increasingly competitive disposal of these managed surpluses contributed to the long-term undermining of peripheral agricultures. Ultimately this process was rooted in the world-historical dynamic of settler agricultural capitalism, which rivaled tropical exports via technological substitution (e.g., corn syrup for sugar) and domestic agricultures via trade substitution (Friedmann 1993).

The food regime perspective informs the chapters written by Friedmann, Thompson and Cowan, Talbot, and Ufkes. Thompson and Cowan periodize their discussion of industrial (canned) foods as midwife to the organization of the world-economy and the transition from preserved to wholly manufactured foods characteristic of the postwar food regime. Friedmann ponders the political implications of the breakdown of the postwar food regime as symptomatic of a more fundamental transformation in political organization and social life. Talbot identifies, in the tropical commodity complex, a counterforce to globalization in the specifically Third World state-capital alliance against transnational food company control of world agriculture. And Ufkes locates "industrial greening" in Singapore within the global disorder attending the breakdown of Bretton Woods and the postwar food regime, intensifying organization of food production and consumption relations on a global scale.

The third thread is that of global commodity chains. Commodity chain analysis, which facilitates mapping of the world division of labor (Hopkins and Wallerstein 1986), has been elaborated in the sixteenth annual PEWS conference volume (Gereffi and Korzeniewicz 1994). This more directly empirical line of inquiry lends itself to short-term analysis. It traces the links among labor and production processes geared toward the creation of a finished product. While such links have always operated globally, they have arguably become more complex and fluid in the present world conjuncture, destabilizing conditions for producers and states, especially in the Third World (Raynolds et al. 1993). The existence of global commodity chains privileges transnational companies, whose operations more evidently pivot on a global labor force (Frobel, Heinrichs, and Kreye, 1979), and on the infinite possibilities of global sourcing of components. Chapter 6 illustrates the corporate reorganization of the carrageenan commodity chain, and chapters 7 and 9 illustrate the limits and possibilities of state regulation of global commodity chains—either as strategies of world-economic positioning, or of containing transnational corporate power in the interests of regulation of local labor markets or environments.

CONCLUSION

The collective emphasis of this volume is that we are not studying agricultural and food systems in different parts of the world so much as coming to grips with how the world is in fact constituted and reconstituted around such systems. And if agriculture and food have languished in the shadow of manufacturing and finance in world-economic accounts of power and historical transformations, this is evidently no longer the case. As critical inputs to the wage-relation, and/ or as the basis for local survival in material and cultural senses, agriculture and food are central to the tensions in the process of globalization. In a world in which the limits of the industrial paradigm are fast becoming apparent, we are rediscovering the historic fact that control of land and food has been fundamental to the political equation—within and among states on the one hand, and through the construction and reconstruction of diets on the other hand. The passage, through this century, from agrarian question to food/green question now appears to be undergoing a radical reversal. That is, global movements resisting the corporate-sponsored self-regulated market ideal are seeking to reintegrate these historically separated questions.

NOTE

The conference from which this edited volume derives was sponsored by the New York State Agricultural Experiment Station, the Cornell University Rural Sociology Department, and the Cornell University International Political Economy Program. These organizations also kindly subsidized publication of a paperback edition. The editor wishes to thank Kevin Sharp of Cornell University Conference Services for his generous assistance in organizing the original conference, as well as the authors for their responsiveness.

REFERENCES

Arrighi, Giovanni. 1990. "The Developmentalist Illusion: A Reconceptualization of the Semiperiphery." In *Semiperipheral States in the World Economy*, edited by William G. Martin, 18–25. Westport, Conn.: Greenwood Press.

Block, Fred. 1977. *The Origins of International Economic Disorder*. Berkeley: University of California Press.

Braudel, Fernand. 1972. "History and the Social Sciences." In *Economy and Society in Early Modern Europe*, edited by Peter Burke, 11–40. London: Allen & Unwin.

Buttel, Frederick H., and Philip McMichael. 1988. "Sociology and Rural History: Summary and Critique." *Social Science History* 12, 2: 93–120.

Cheru, Fantu. 1989. *The Silent Revolution in Africa: Debt, Development and Democracy*. London: Zed Books.

Cleaver, Harry. 1977. "Food, Famine and the International Crisis." *Zerowork* 2, 7–70.

Friedmann, Harriet. 1978. "World Market, State and Family Farm: Social Bases of Household Production in the Era of Wage-Labor." *Comparative Studies in Society and History* 20, 4: 545–86.

———. 1993. "The Political Economy of Food: A Global Crisis." *New Left Review* 197: 29–58.

Friedmann, Harriet, and Philip McMichael. 1989. "Agriculture and the State System: The Rise and Decline of National Agricultures, 1870 to the Present." *Sociologia Ruralis* 29, 2: 93–117.

Frobel, Folker, Jürgen Heinrichs, and Otto Kreye. 1979. *The New International Division of Labor*. New York: Cambridge University Press.

Gereffi, Gary, and Miguel Korzeniewicz, eds. 1994. *Commodity Chains and Global Capitalism*. Westport, Conn.: Greenwood Press.

Hopkins, Terence K., and Immanuel Wallerstein. 1986. "Commodity Chains in the World-Economy prior to 1800." *Review* 10, 1: 157–70.

McMichael, Philip. 1984. *Settlers and the Agrarian Question: The Foundations of Capitalism in Colonial Australia*. London: Cambridge University Press.

Polanyi, Karl. 1957. *The Great Transformation: The Political and Economic Origins of our Time*. Boston: Beacon.

Raynolds, Laura, David Myhre, Philip McMichael, Viviana Carro-Figueroa, and Frederick H. Buttel. 1993. "The 'New' Internationalization of Agriculture: A Reformation." *World Development* 21, 7: 1101–21.

Sayer, Derek. 1991. *Capitalism and Modernity. An Excursus on Marx and Weber*. London: Routledge.

Taylor, Peter. 1987. ''The Poverty of International Comparisons: Some Methodological Lessons from World-System Analysis.'' *Studies in Comparative International Development* 22: 12–39.

Wallerstein, Immanuel. 1974. *The Modern World-System.* Vol. 1, *Capitalist Agriculture and the Origin of the European World-Economy in the Sixteenth Century.* New York: Academic Press.

———. 1984. *Historical Capitalism.* London: Verso.

———. 1991. *Unthinking Social Science: The Limits of Nineteenth Century Paradigms.* Cambridge, England: Polity Press.

Watkins, Kevin. 1991. ''Agriculture and Food Security in the GATT Uruguay Round.'' *Review of African Political Economy* 50: 38–50.

PART I

Food in World-Historical Perspective

1

Food and Its Relationship to Concepts of Power

Sidney Mintz

Grace to Be Said at the Supermarket

That God of ours, the Great Geometer,
Does something for us here, where He hath put
   (if you want to put it that way) things in shape,
Compressing the little lambs in orderly cubes,
Making the roast a decent cylinder,
Fairing the ellipsoid of a ham,
Getting the luncheon meat anonymous
In squares and oblongs with the edges beleved
Or rounded (streamlined, maybe, for greater speed).

Praise Him, He hath conferred aesthetic distance
Upon our appetites, and on the bloody
Mess of our birthright, our unseemly need,
Imposed significant form. Through Him the brutes
Enter the pure Euclidean kingdom of number,
Free of their bulging and blood-swollen lives
They come to us holy, in cellophane
Transparencies, in the mystical body,

That we may look unflinchingly on death
As the greatest good, like a philosopher should.
                                        —Howard Nemerov

INTRODUCTION

This chapter presents my view of how a society learns to consume food differently: eating more food (or less), eating different food, differently prepared, in different contexts, with the social (and perhaps the nutritive) purpose of the consumption itself revised or modified. My aims are roughly to block out, on the one hand, the sorts of constraint that define the situations within which people accept the need to change their food habits and, on the other, to illustrate how they cope with change cognitively and symbolically, by constructing new frames of consumption with modified meanings.

The use and application of power enters into many such instances of change. Where this power originates, how it is applied and with what ends, and in what manner people undertake to deal with it are all part of what happens. Even though they are of immense importance to the world's future, we do not understand these processes at all well. Furthermore, much of the research on changing food habits does not address the cultural aspects of such changes. Group values and past practices can figure significantly in what changes are made, how many and how fast. Thus culinary history enters into the success and failure of new applications of power in the sphere of food and eating, but not in readily understood or carefully studied ways. I attempt here to explain, in a preliminary fashion, my own ideas of how power serves to advance (or retard) changes in food habits.

SUGAR, TEA, AND THE BRITISH WORKING CLASS

The heightened use of tea, sugar, tobacco, and a few other substances, which came to typify the use and spending habits of the eighteenth-century British working class, probably provides us with the first instance in history of the *mass* consumption of imported food staples (Mintz 1985). Although the hope in *Sweetness and Power* (Mintz 1985) was to be able to explain the peculiar attraction these novelties had for new consumers, the argument remained incomplete in part because it was impossible to locate and isolate some specific single cause for this new consumption. Many explanations had turned up in the literature; none seemed particularly convincing. Two historians, sniffing the air anew, settled on ''the quest for respectability'' as *the* cause (Austen and Smith 1990; Smith 1992). Respectability, concrete and specific though it sounds, does not take us far. We still do not really know why so many English people so rapidly became such eager consumers of sugar and tea, for example. The term ''respectability'' can be an umbrella for such themes as hospitality, generosity, propriety, sobriety, social rivalry, and much else (Mintz 1993). The unanswered (and perhaps unanswerable) question persists if what we aim at explaining is the peculiar power of a *specific* food (or even some category of foods) over consciousness and will. Among possible factors influencing the British adoption of sugar, I earlier noted the powerful stimulant contained in tea and other new beverages, coffee and chocolate, with which sugar was consumed; the malnutrition of the British working classes at the time, such that the caloric contribution of sugar would matter unconsciously as well as consciously; the apparently universal predisposition of the human species toward the sweet taste;

the readiness of people in most (if not all) societies to emulate their "superiors" if permitted; the possible significance of the element of novelty; and the usefulness of tobacco and the stimulant beverages in easing the industrial work day (Mintz 1985). Faced with such a list, it becomes harder to talk about the relationship between food and power in terms of some single specific food.

The emergence of British sugar-eating and tea-drinking took place against a background of overseas expansion and colonial conquest: mounting commerce in enslaved Africans and growing plantations in the colonies and increasing industrialization, dislodgement of rural populations, and urbanization at home. Sugar itself, earlier a rare and precious imported medicine and spice, became cheaper (at first rapidly, then more gradually), while it did, the uses to which it might be put proliferated. Sugar's increasing availability facilitated the increase in contexts within which it might be used.

Once sugar began to be consumed by those of modest income, its application increased swiftly. It entered into the rhythms of daily life particularly in its association with the three new stimulant beverages (in Britain, tea soon emerged as the most successful). Much later, and by a series of successive steps, sugar became important in its own right—that is to say, other than in association with these beverages.

MEANING

In studying materials dealing with home and work conditions in Britain in relation to sugar and other substances, it was useful to separate the broad changes in background that made access to sugar easier, on the one hand, and the circumstances of daily domestic life and work, within which consumers installed sugar in their everyday routines, on the other. On this basis, two terms were proposed to simplify discussion. The daily life conditions of consumption had to do with "inside meaning"; the environing economic, social, political (even military) conditions, with "outside meaning."

Inside meaning arises when the grand changes connected with outside meaning are already under way. These grand changes ultimately set the outer boundaries for determining hours of work, places of work, mealtimes, buying power, child care, spacing of leisure, and the arrangement of time in relation to the expenditure of human energy. In spite of their significance for everyday life, they originate outside that sphere and on a wholly different level of social action. In consequence of these changes, however, individuals, families, and social groups must busily integrate what are newly acquired behaviors into daily or weekly practice, thereby turning the unfamiliar into the familiar, imparting additional meaning to the material world, and employing and creating significance at the most humble levels. This is what happened to tea-drinking, once people had tasted tea and were learning to drink it regularly, and what happened to pipe-smoking, once tobacco had been tried and was liked. People alter the micro-conditions as much as they can and according to their emerging preferences—the where, when, how, with whom, with what, and why—thereby changing what the things in question signify, what they *mean* to the users. New behaviors are superimposed upon older behaviors; some behavioral features are retained, others forgone. New patterns replace older ones.

This happens, however, within the widest constraints that outside meaning permits. The processes that endow behavior with inside meaning unfold in relation to what I label "grand changes." But, of course, for the participants the micro-conditions themselves are, or become, grand—for it is out of them that the routines of daily life are fashioned. This interior embedding of significance in the activity of daily life, with its specific associations (including affective associations) for the actors, is perhaps what anthropologists have in mind when they talk about meaning in culture.

Some of us tend to be inordinately moved by the power of our species to invest life with meaning on this intimate, immediate, and homely level. It is essential to stress the remarkable—even distinctive—capacity of our species to construct, and act in terms of, symbols. But in the case of the large, complex societies with which we deal today, it is at least as important to complete understanding to keep in mind that larger institutional subsystems usually set the terms against which these meanings in culture are silhouetted. In daily practice, for example, job opportunities tell people when they can eat and how long they can take to do it; to a noticeable extent, they also therefore tell people what they can eat, where, and with whom. Individuals are thus presented with a series of situations within which they may begin to make meaningful constructions for themselves, as long as such constructions do not violate the outer situational boundaries created for them. But the job opportunities are determined by forces that transcend the means and wills of those who become the employees—as anyone who has lost a job recently knows.

In contrast to inside meaning, it is those larger forces expressed in particular subsystems, together with the state, that have to do with what I meant by the term "outside meaning." Thus, outside meaning refers to the wider social significance of those changes effectuated by institutions and groups whose reach and power transcend both individuals and local communities: those who staff and manage the larger economic and political institutions and make them operate.

In the case of the history of sugar in Britain, it was the servants of the imperial political and economic system who carved out the West Indian colonies and gave them governments; saw to the successful—immense and centuries-long—importation of enslaved Africans to the islands; bequeathed land wrested from the indigenes to the first settlers; financed and managed the ever-rising importation of tropical goods to Britain, including chocolate, coffee, cotton, and tobacco, as well as sugar, rum, molasses, tea, and much else; and levied taxes at all levels of society to benefit its servants and the state. It will soon become clear that these background arrangements of conditions against which inside meaning can take on its characteristic shape—what I call outside meaning—are cognate with what Eric Wolf has labeled "structural power" (Wolf 1990, 586–87).

To have used the word "meaning," rather than "power," in the first of my labels ("outside meaning") may have been somewhat misleading, but there was a reason for it at the time (Mintz 1985). For a decade or two, anthropologists have gradually been abandoning an older interest in causation—in trying to explain why *this* happened, rather than *that*—in order to make analyses of events in terms of what is thought to be their meaning. We are told that such a shift in emphasis has had the salutary effect of bringing the ethnographer into view and of demystifying both the ethnographer and the fieldwork situation (though neither is a genuinely new idea).

But the stress on meaning has also led us away from seeking to explain what happens (or happened) in the course of human events. We know that particular events often *mean* different things to different persons or groups in the same society. The slave trade and slavery "meant" that the British factory and farm workers would get their sugar; however, the meaning of slavery and the slave trade to plantation owners, bankers, and the Colonial Office was entirely different. (One need hardly add that, for the slaves and their descendants, it also "meant" quite different things.) For anthropologists still interested in how things happen and the consequences of events, there has to be a way to distinguish among different meanings, and different sorts of meaning, in order to continue to study causation. The abstract system we call a culture, as well as the abstract system of meaning thought to typify the members of the society who share that culture, are neither simple coefficients of each other, nor two sides of one coin, nor merely the active and passive aspects of one system. To treat them as if they were is to bypass the complex nature of any society and to impute to its members a homogeneity of value and intentions they almost certainly lack. Outside meaning was a term invented to avoid the imputation of any such homogeneity.

As for inside meaning, the term "meaning" is entirely appropriate here, I think. Those who create such inside meaning do so by imparting significance to their own acts and the acts of those around them, in the fashion in which human beings have been giving their behavior such social significance as long as they have been human. The gradual emergence of a food pattern called "high tea" among working-class Britons was the work of those who eventually came to take this meal regularly; it was they who created the pattern. But they did so inside the constraints of work and income and their own available energy, constraints over which they themselves had hardly any control at all.

The connection between outside and inside meaning can be exemplified with a more modern case than that of sugar and tea in eighteenth-century Britain. But before looking at this case, we need to take note of a general paradox having to do with the whole issue of food and food preferences. On the one hand, food preferences, once established, are usually deeply resistant to change. We cannot easily imagine the Chinese people giving up rice to eat white bread, or the

Russian people black bread to eat maize. Such deeply cherished tastes are rooted in underlying economic and social conditions, and they are surely far more than simply nutritive. But they must also be viewed in terms of the equally telling fact that *some* preferences, even in diet, turn out in fact to be quite readily surrendered. To be sure, it is far more common to add new foods to one's diet than it is to forgo old and familiar ones. The readiness of North Americans to become eaters of *sushi*, which surely could not have been predicted in 1941—and not only for political reasons—is an apt example of an unexpected, even unpredictable adding-on. Somewhat more interesting in the present argument is the gradual decline in the consumption of complex carbohydrates by North Americans over the past seventy-five years, which has meant not just the addition of new foods, but also a palpable decline in the consumption of certain once-prized old ones. In any event, these addings-on and gradual eliminations are often hard to explain for they proceed against substantial, persisting stability of diet at the same time.

We do not understand at all well why it can be claimed both that people cling tenaciously to familiar old foods, yet readily replace some of them with others. Hence situations of rapid change in food habits deserve a much closer look than they have received. We need to know far better than we do now why some food habits change easily and swiftly, while others are remarkably enduring. We are inclined to view this contrast as between basic or essential foods on the one hand, and less important or peripheral foods on the other. But this is not adequate to explain all particular cases of rapid change. When much else is changing, food habits may change, too, and such changes are often unpredictable. Where and how power enters into these processes of change, projected in part against continuing stability, is not always apparent.

A CASE

Here, then, is one example of how such changes may work to affect food choices. This large-scale, general case has to do with war. War is probably the single most powerful instrument of dietary change in human experience. In time of war, both civilians and soldiers are regimented—in modern times, more even than before. There can occur at the same time terrible disorganization and (some would say) terrible organization. Food resources are mobilized, along with other sorts of resources. Large numbers of persons are assembled to do things together—ultimately, to kill together. While learning how, they must eat together. Armies travel on their stomachs; generals—and now economists and nutritionists—decide what to put in them. They must do so while depending upon the national economy and those who run it to supply them with what they prescribe or, rather, they prescribe what they are told they can rely upon having.

During World War II, upwards of fifteen million Americans were brought together in uniform, many millions more in mufti. The service people ate together in large camps. They ate what they were given; what they were given was decided by power holders who functioned outside the army and outside their direct experience.

Among other things, service personnel were given meat twenty-one times a week; even the Friday dinner had an alternate meat course (though it was usually cold cuts). For most soldiers (but only irregularly under combat conditions), never before had so much meat been thrust before them. They were also given vast quantities of coffee and sweets of all sorts; there were sugar bowls on every table and twice a day, without fail, the meal ended with dessert. Soldiers were also given free cigarettes in the pay line each month (by a ruse that lined the pockets of some noncommissioned officers). Though the food habits of the civilians may not have been so radically altered, certain things did happen, about which much is known. Civilians got too little meat; the wartime media were full of stories and jokes about romancing the butcher. They got too little sugar, too little coffee, and too little tobacco. Their food habits were also being radically affected. Hence North American food preferences—though "preferences" may seem slightly misleading, under the circumstances—were significantly reshaped by the war experience.

Among the things that soldiers and civilians were *not* given was Coca Cola, but it was carefully arranged that they could buy it. George Catlett Marshall, chief of staff during World War II, was a southerner. Soon after Pearl Harbor General Marshall advised all of his commanders and general officers to request Coca Cola bottling plants in order to get the product to the front. By his letter Marshall gave Coca Cola the same status in the wartime economy as that occupied by food and munitions. Coca Cola was thus spared sugar rationing. In all, sixty-four Coca Cola plants were established in allied theaters of war, including the Pacific theater, North Africa, Australia, and elsewhere. The Coca Cola Company was asked by the armed forces to supply technicians to run the production; 148 bottling plant technicians were sent; three were even killed in theaters of war during World War II (Louis and Yazijian 1980).

In the light of Coca Cola's status by the time the war ended, it is noteworthy that, before the war, Coke was not only not much of an international drink—it was not even much of a national drink. Though Coke had traveled early in its career to Cuba, it was still principally a U.S. beverage, consumed mainly in the South. It was sold in some foreign countries, but it certainly was not well known internationally. In most places in the United States where Coke was drunk, high school students laced it with Southern Comfort whiskey so that they could get publicly (yet covertly) intoxicated at the senior prom. Indeed, it is probable that most people outside the South never drank Coke, but "mixed" it instead. During the war, the fact that the U.S. professional officer corps was largely southern may have played a role in this story, as well.

How outside meaning was associated with the *spread* of Coca Cola is easy to discern. The rapid proliferation of Coca Cola bottling plants in allied theaters of war speaks directly to the issue. Power over labor and resources employed in the production of food undergirded the unhampered operation of the corporate system, closely coordinated in this instance with the will of the state. Even in times of politico-military crisis—some might say particularly in such times—corporate power neatly integrated with the state bureaucracy firmly underwrites the successful execution of broader societal tasks. At such moments, the power of the state itself seems far less irksome to corporate America. The deployment of resources for food production is linked to conceptions of consumer choice as well. But in this instance the choices were managed in a specific fashion: 95 percent of all soft drinks sold on American bases during the war were products of the Coca Cola Company. There was choice, but one company was accorded the right to specify its limits.

In contrast to outside meaning, inside meaning in a case of this kind has to do with what foods come to mean to those who consume them. The symbolism connected with Coca Cola, as it took on its national stature during the war, was utterly astonishing: The first bottle to come ashore at Anzio, for instance, was shared by nineteen GIs. It was not unusual to find in the letters that servicemen wrote home the assertion that they were fighting for the right to drink Coca Cola. The inside meaning of Coca Cola is certainly revealed in the emotions of a soldier who fights—among other things—"as much to help keep the custom of drinking Cokes as I am to help preserve the millions of other benefits our country blesses its citizens with"—to quote from one of many such references to Coke in the censored mail of wartime. Thus it was that Coca Cola was enabled to become a symbol—a veritable national symbol—among the warrior youth of the 1940s generation.

War, then, is a setting in which the exercise of the power behind outside meaning readily applies. Such examples do not have to do with the intrinsic nutritive significance of food. They help to explain, rather, how outside processes serve to impose many of the conditions within which inside meaning can take shape and manifest itself.

CONCLUSIONS

In his lecture to the American Anthropological Association annual meetings some years ago, Eric Wolf enumerated four sorts of power (Wolf 1990). By Wolf's reckoning there is, first of all, personal power, of a sort comparable to charisma. Second, there is the power of persuasion, by means of which one person exacts conformance of some kind from another. Third, and on a broader canvas, there is the "power that controls the settings in which people may show forth their potentialities and interact with others" (1990, 586). This "tactical or organizational power" is "useful for understanding how 'operating units' circumscribe the actions of others within determinate settings." Tactical power can be used, for example, by organized business entities, such as multinational corporations, banks, and conglomerates. The exercise of such power is tightly linked to the current argument. Even more important is Wolf's last category:

a fourth mode of power, power that not only operates within settings or domains but that also organizes and orchestrates the settings themselves, and that specifies the distribution and direction of energy flows. I think that this is the kind of power that Marx addressed in speaking about the power of capital to harness and allocate labor power. . . . I want to use it as power that structures the political economy. I will refer to this kind of power as structural power. This term rephrases the older notion of "the social relations of production," and is intended to emphasize power to deploy and allocate social labor. These governing relations do not come into view when you think of power primarily in interactional terms. Structural power shapes the field of action so as to render some kinds of behavior possible, while making others less possible or impossible. (1990, 586–87)

When this perspective is applied to the subject of food habits, it is easy to see how structural and tactical (or organizational) power undergird the institutional frameworks that set the terms by which people get food, maintain or change their eating habits, and either perpetuate their eating arrangements and the associated meaning or build new systems, with new meanings, into those arrangements.

All living organisms are faced with an imperious necessity: not to eat is to die. But beyond this, foods have meanings that transcend their nutritive role. Just as our species seems always to have made food carry symbolic loads far heavier than those of simple nutrition, so, too, the symbolism seems ready to spill over into even wider fields of meaning. The place of rice in Japanese culture, of bread in the West, of maize to many Native American peoples—these significations clearly surmount any literal nutritive significance the foods themselves might have.

It might seem acceptable to say, then, that food exercises "power" over people in terms of what it means to them. But that is *not* the sort of power with which this author is dealing, and it is important to be clear in this regard. The material world is invested with meaning; because people act in terms of understood meanings, meaning can be said to effectuate behaviors of certain kinds. Power and meaning are always connected. "Power is . . . never external to signification," Wolf writes, "it inhabits meaning and is its champion in stabilization and defense" (1990, 593). But the symbolic power of foods, like the symbolic power of dress or coiffure, is different from (even if related in some manner to) the tactical and structural power that sets the outermost terms for the creation of meaning. The power within outside meaning sets terms for the creation of inside, or symbolic, meaning.

Turn again to the words of that earnest GI who fought to preserve his right to drink Coke. There is no question about inside meaning in this instance. Such inside meaning is linked to outside meaning because what Coke *means* is coefficient with its history as a commodity, with the steps taken to ensure its availability, with the history of those very decisions by which Coke could become the purchased soft drink, the tie to home, the exciter of nostalgia, the *symbol* of America. What I have called outside meaning and inside meaning are clearly linked in Coke's story, but they are clearly different from each other, and they do not stand in any simple relationship.

In his own work, Wolf has set apart the issue of meaning from the issue of power, but he sees them as inextricably connected: "Meanings are not imprinted into things by nature; they are developed and imposed by human beings. Several things follow from this. The ability to bestow meanings—to 'name' things, acts and ideas—is a source of power" (1982, 388).

As this chapter attempts to suggest, the ability to "supply" things, in the broadest sense, is also a vital source of power, not only because it may include some ability to bestow meaning, but also because meaning coalesces around certain relationships. Objects, ideas, and persons take on a patterned structural unity in the creation of ritual, as happened, for example, when high tea became a working-class eating custom. But it was the purveyors of the foods, the givers of employment, the servants of the state who exercised the power that made the foods available.

If we return briefly to the case of sugar in eighteenth-century Britain, we may inquire of the material to what extent the creators of the background conditions can be said to have set the precise terms for the emergence of inside meaning. Emulation, for example, played some role in increasing and in shaping use; so, probably, did medical advice. The conditions under which landless people worked were determined by others: the hours when they might eat or rest, where they took their food, how they got to and from work. At the level of daily life, the customary practices that working people developed in order to deal with the newly emerging industrial society in which they found themselves were answers, or "solutions" to conditions over which they had no real control. In these ways, outside and inside meanings are linked through the conditions created and presented to potential consumers by those who supply what is to be consumed.

This chapter has aimed at clarifying these questions, not at answering them seriously. What is needed is a concerted effort to study the various ways in which stable food habits can be called into question. We may also ask ourselves *why* they are called into question. Some answers may have to do with poor nutrition, overeating, or inordinately expensive cuisine, relative to available resources. But other answers may have little or nothing to do with health or economy, even though people are being subjected to intense pressures to forgo some parts of their diet in favor of different foods. At times, as has been suggested here, large-scale structural changes, such as war and migration, may change the rules of the game, so to speak, compelling people to reorder their categories of meaning in new ways, and to eat (and drink) differently. How this is done, and why it succeeds, badly needs to be understood. So, too, do all of the means used to persuade people that what they are eating now should be replaced with something else. I think that it is within anthropology's capabilities to confront these issues solidly, but it has not done so. Until it tries to accomplish this, it will not be able to contribute fully what it can to our understanding of the world food problem.

NOTE

This chapter was delivered in draft at the seventeenth annual PEWS meeting in Ithaca, New York, in April 1993. The author is grateful to Gillian Feeley-Harnik and Eric Wolf for comments on an early draft; to Immanuel Wallerstein and Harriet Friedmann for their commentary at the meeting; to Dale Tomich, Phil McMichael, and John Walton for valuable criticisms of a later draft; and to Jackie Mintz for repeated readings and advice. Unfortunately, the author has been able to use only a small part of their collective wisdom; no one but he is responsible for any persisting errors and excesses.

REFERENCES

Austen, Ralph, and Woodruff Smith. 1990. "Private Tooth Decay as Public Economic Virtue: The Slave-Sugar Triangle, Consumerism, and European Industrialization." *Social Science History* 14, 1: 95–115.

Louis, J. C., and Harvey Z. Yazijian. 1980. *The Cola Wars.* New York: Everest House.

Mintz, S. 1985. *Sweetness and Power.* New York: Viking–Penguin.

———. 1993. "The Changing Roles of Food in the Study of Consumption." In *Consumption and the World of Goods*, edited by J. Brewer and R. Porter, 261–73. London: Routledge.

Smith, Woodruff. 1992. "Complications of the Commonplace: Tea, Sugar and Imperialism." *Journal of Interdisciplinary History* 23, 2: 259–78.

Wolf, Eric R. 1982. *Europe and the People without History.* Berkeley: University of California Press.

———. 1990. "Facing Power: Old Insights, New Questions." *American Anthropologist* 92, 3: 586–96.

2

Food Politics: New Dangers, New Possibilities

Harriet Friedmann

Not for the first time in history, food and agriculture are focal points of social and international conflict. Yet we are little prepared to understand the changes upon us. During the postwar boom, the celebration of industrial production and consumption pushed food and agriculture to the margins of popular and scientific interest. Yet common sense still serves as a starting point for recovery. Food and agriculture are enduring moments of social organization. No matter how complicated world trade and finance become, we all have to eat. No matter how sophisticated biotechnologies become, the plants and animals we eat ultimately come from the earth.

Certain theories of political economy retain the wisdom in this common sense. Karl Polanyi, in particular, built a theory on the premise that land and labor, the human and natural substance of society, are commodified at our peril. To commodify land and labor, according to Polanyi, is to attempt to disembed the universal conditions of society from the conscious and complex relations and practices of human beings. Full commodification of land and labor is a utopian project based on belief in the possibility and desirability of a self-regulating market. Because the project is utopian, eventually attempts to expand and deepen the scope of markets cause serious damage to human beings and our habitats. Therefore, periodically, we respond by measures to protect people and land. Since the rise of industrial capitalism, history can be seen as a pendulum moving between self-regulation by markets and self-protection by society.

Food becomes important as the pendulum approaches either pole: Self-regulation by the market wreaks havoc on land and labor and, therefore, on the long-term bases of the economy; however, self-protection against the market limits the maneuver and, therefore, the short-term profitability of capital. When Polanyi wrote in the midst of World War II, the pendulum had swung back to protection in reaction to the Great Depression. Until they collapsed, markets had been given sufficient rein between 1870 and 1930 to concentrate commercial food production in the North American plains and several other places. The result, after decades of agricultural relocation from Europe to settler regions, was a collapse of international commerce and finance—the Great Depression—and an ecological crisis—the dust storms that blew away soil and settlers after only one or two generations.

Among the various experiments in self-protection in Polanyi's time (which included fascism and Soviet communism), the one that eventually dominated was the U.S. New Deal. This centered on price supports for agricultural commodities, supplemented by a variety of measures including import controls and export subsidies to dispose of government-held surplus stocks. These are the policies, adopted in modified forms by most significant capitalist food-trading countries in the decades after World War II, that eventually led to international economic conflict in the 1980s and 1990s.

To judge by the rhetorical celebration of "markets" in all nations and in international fora, and by the practices of austerity, privatization, deregulation, and the like, the pendulum is swinging back. A new era is being constructed, in which people and the earth are forced to "adjust" to the "market," and it is the markets, not people, that require freedom. The swing began a decade ago in the Third World and former socialist countries under the debt regime imposed by the International Monetary Fund (IMF) and other internal agencies. Now it includes the core.

The question arises: Are we destined to swing between the poles of too much market and too much regulation? Or is it possible for human communities to regulate economic life in ways that enhance both their cultural life and their natural habitats? Within the highly state-regulated agriculture of the past fifty years, commodification of land and labor has proceeded so deeply and so widely that most of the people of the world depend on food markets and therefore on commercial agriculture. Moreover, many transnational food corporations, which grew up within the framework of national regulation, now experience that same national regulation as a constraint to further growth. The combined weight of economic power and market dependence support the swing back to self-regulated markets. If all goes well, corporate reorganization of global markets will allow profits to grow for another several decades. However, the human and ecological consequences will eventually bear heavily on world politics—as they did during the Great Depression and the world wars which dominated the first part of this century. It may already have begun in the wars and famines spotting the globe.

Is there a happy medium? It is worth considering, like Goldilocks, that if the world market is too hot and state agricultural regulation is too cold, perhaps there is a way to organize *food economies* that is just right. I would like to explore this question by outlining the main shifts in power, agency, issues, and social movements that have taken place in the framework of the food regime created after World War II.

First, I shall briefly illustrate my starting point: Although dramatic changes in diets and cuisines began at least four hundred years ago, with colonial movements of people and plants across the globe, nonetheless the character and pace of change altered radically when the agency shifted from colonial states and merchant companies to national states and transnational corporations.

Second, turning to agro-food production, in the decades after World War II, power shifted away from farmers who initiated the current agricultural policies. It shifted strategically toward corporations and numerically toward consumers. At the same time, work in the agro-food economy shifted away from self-employed farmers to waged workers in manufacturing and services. This shift entailed a change in the gender and ethnic, as well as class, composition of the agro-food sector. Farmers, despite the crucial uncounted labor of family members, were statistically and politically visible as mainly native born or European origin men; industrial and service workers were recruited largely among women, teens, and ethnic minorities of both genders. Work also shifted from paid to unpaid work, mostly by women, of shopping, preparing, and serving meals in increasingly commodified households (Friedmann 1990; Glazer 1990).

Third, changes in diets and cuisines (Weismantel 1988: 87) result from shifting balances between coercion and consent in the constitution and reconstitution of desire for specific foods. These balances come into play everywhere with the commodification of labor, land, and daily life.

Finally, I will suggest an alternative to the dilemma between markets versus subsistence: It is not markets or regulation that matters, as dominance of either is either too hot or too cold to exist in real life. Instead, what matters is the scale and embeddedness of food economies in human communities adapted to their habitats.

COLONIAL POWER AND CORPORATE POWER

Diets and agriculture changed nearly everywhere during the three centuries of colonial conquest and trade which preceded the industrial revolution. If colonial empires moved people and plants around on an unprecedented scale, then industrial food marked the next stage of dietary change (Goody 1982). Although industrial food originated with military provisions and colonial products, its present incarnation is the creature of a new social institution, the capitalist enterprise. Those enterprises now bear complicated relations to their own countries of origin. Their operations have expanded and contracted as states have contested, negotiated, and renegotiated rules of world investment and commerce. However, the histories of dietary change, including the shifting role of industrial and corporate food, continues to bear the cultural legacy of contested power and property.

What is distinctive about dietary and agrarian change in today's world? Both have always changed, of course, through conquest, migration, trade, and plunder. People have repeatedly adopted new plants and animals into their farming, cultivated their palates, and lost contact with regions of former habitation or trade. The idea of tradition is dangerous if it fosters the image of timeless practices by stable communities. The world has been disrupted and integrated on a global scale for centuries, and people, plants, and practices have been relocated and reshaped many times. In search of spices from Asia, Europeans conquered much of the globe and, by the eighteenth century, had forced or induced laborers on plantations and farms to produce sugar, coffee, cocoa, tea, and opium in sufficient quantities to stimulate, soothe, and compensate the overworked, overcrowded, impoverished workers gathered in factories, fields, and cities by the industrial transformation of life in Europe (Mintz 1979).

Merchants and colonists brought to the Americas Old World plants, such as barley and wheat, and livestock, such as chickens and cattle (Super 1985). Eventually, one of those crops, barley, became so deeply incorporated into the agrarian and culinary practices of the Zumbagua, self-provisioning Quechua speakers in the Ecuadorian Andes, that they now identify their cultural distinctiveness via a barley-based staple food—"we who eat ma'chica" (Weismantel 1988, 181).

The slave trade brought New World plants to Africa which became agricultural and dietary staples, notably maize, yams, and manoic. Slaves captured in Africa learned to eat salt cod fished by British settlers in the North Atlantic. Transported African slaves in the Caribbean colonies of Europe produced a transplanted Asian crop, sugar, to sweeten English workers' tea—in turn, a Chinese crop grown in colonial plantations in India and Ceylon. To close the circle, colonial rule brought new ingredients to old as well as new colonies. The dishes of India, China, Thailand, and Indonesia adapted Mexican chilis, just as Asian indentured laborers brought mangoes and curries to the cuisine of the Caribbean.

Until the middle of the twentieth century, most dietary changes occurred locally. Even when plants from distant lands became essential crops and ingredients, they were adopted into regional food economies. Even if they changed dramatically (for instance, try to imagine the gardens and cuisines of southern Europe and North Africa without the New World tomato), men and women learned to grow, prepare, and enjoy new plants by embedding them within their daily and seasonal rounds of life.

These adoptions were not necessarily spontaneous or even voluntary. For instance, the potato, the New World crop that most decisively changed the staple foods of Europe, had a complex and ambiguous relation to the social transformations that created new classes, races, and hegemonic states. Long an exotic food, even an aphrodisiac, among the privileged ranks of England (Salaman 1985, 424–25), the potato gained its first foothold as a staple of the masses in Ireland. It did so in conditions of forced regression of both industry and agriculture under the restrictions imposed by English conquerors against Irish Catholics. Salaman describes the circumstances at the end of the seventeenth century:

The workers displaced from industry had begun to trek back to the countryside to swell the growing army of the new poor. Many, who formerly had been large landowners, cultivators or graziers, had now, by reason of the confiscations, and the penal laws of the last half of the century, returned to the land to take their place alongside the peasantry they had in former times contemned [*sic*]. Already the poverty-ridden cottier class, driven by economic and political forces over which it had not the slightest control, had adopted the potato as its main source of food. From now on, this same prolific root was called on to maintain the greater part of the entire Irish nation. In the beginning of the century . . . the potato, a new type of food plant, a complete stranger to Europe, had reached the Irish people at a critical moment, without credentials or reclame, with no accredited representative to blaze its path, or acclaim its merits. Its arrival coincided with that moment in Irish history when destruction of the material basis of existence had reached its climax, and the amenities of life their nadir. The people, rich and poor alike, were in dire need of a foodstuff which would withstand the vagaries of nature and the malignity of man. (1985, 243)

A century later, when the governing ranks, abetted by philanthropists, attempted to introduce the potato as a new staple food into England itself, the struggle was already engaged over the standard of living of the new working classes. By that time, diet, like all conditions of life, had deteriorated for rural and urban workers alike, and they clung to the ideal of the white loaf. Attempts to introduce potato flour into bread, and later potatoes as such, were resisted by many means, including food riots. This was so despite the famine conditions brought by crop failures and then by the Napoleonic Wars. By that time, too, potatoes, like other conditions of life, were associated by English workers with the Irish, whose immigration and competition for paid jobs were bitterly resented (1985, 503–8). And, indeed, when the potato was finally introduced as a staple of laborers' diets in the early nineteenth century, it was by means of a national economic policy to retreat from the commodification of labor. The state introduced and encouraged allotments, a garden plot in which the "laboring poor" were to cultivate their own food as a supplement to wages (1985, 525).

Remarkably, by the end of the nineteenth century, the potato had become an object of desire and even nostalgia. It has been codified a century later as "traditional" to the regions of Europe: fish and chips, meat and potatoes, steak with "fries," boiled potatoes with dill, and of course vodka. That change in desire, and parallel changes in European farming practices, is a story that began with wars and impoverishment, along with propaganda from private and public figures interested in cheapening the cost of food for the poor and maintaining public order. In this sense, it is just as much and just as little "traditional" to European diets as the barley of the Zumbagua. Each is a story involving webs of power, property, and (with more variation) money, which surrounded those who, by necessity or desire, planted it in their farms, prepared it in their kitchens, and ate it at their meals.

The history of wheat is a different web of conquest, property, commerce, and labor. Wheat farming was an intrinsic part of a comprehensive displacement of indigenous cultures of North and South America, Australia, and New Zealand, by a composite European culture. European settlers brought wheat to these temperate settler colonies with the encouragement of merchants, railway and steamship magnates, and expanding states, both colonial and national. They all had a double purpose: to send wheat back to Europe more cheaply than European farmers could produce it, and to reproduce European dietary culture along with European agriculture in territories seized from native peoples. This was implicated in state-building projects which appropriated lands used by native peoples for different sorts of agriculture or for hunting, and reconstructed the relations between the land and European settlers.

Wheat and cattle, as farming products and as dietary staples, were means of colonization and displacement, a process completed only in the early decades of the twentieth century. That process culminated in the collapse of world markets and ecological crisis in the 1930s. Yet after World War II, only a few decades after the completion of plains settlement in North America, surplus wheat stocks exported to Africa, Asia, and Latin America began a wave of dietary change, and sometimes disastrous attempts at agronomic change, on the part of the people whose lives were transformed rather than displaced.

Now the European meat, wheat, and potatoes diet, which became traditional—with regional variations—over centuries, has within decades become the standard American hamburger (bread and beef) and fried potatoes all over the world. The form is crucial: The meal is defined by MacDonald's and other fast-food restaurants, not by multitudes of cooks experimenting, adapting, and sharing recipes. If the restaurant chain buys cattle and potatoes locally, it places itself between the farmer and the consumer. The farmer responds not to a multitude of experimenting cooks asking and choosing at the market, but to specific, mandatory corporate requirements regarding genetic stock, chemical inputs, and price. This, concretely, is how the market is not what it was when farmers and workers eventually came to love potatoes in late nineteenth-century Europe.

PRODUCTION: CHANGING POWER, CHANGING AGENTS

Apparently, like Goldilocks, we confront choices between food production that is too big or too small: either global organization of land and labor by transnational corporations, or "food first" for households, communities, and nations (Lappé and Collins 1974). Representing the big choice, the editors of The Economist argue for free trade, as the magazine has done since it was founded almost a century and a half ago to oppose British controls on the grain trade (it succeeded when import controls called "Corn Laws" were abolished in 1846). This approach implicitly empowers transnational food corporations, which did not exist in their present form until after World War II. Not only do they embody a transformation in property and power that changes the meaning of free trade, but they also (selectively) integrate agro-food sectors across apparently independent national economies.

By naming current agricultural policies "the new corn laws," *The Economist* anachronistically identifies the farmers of 1993 with the landed gentry of the early nineteenth century. This sleight of hand allows the editors to ignore the real contest between public regulation and private power, as they rail against farmers for somehow managing to hold consumers, governments, even corporations and Third World countries to ransom despite their miniscule and shrinking numbers. Indeed farmers have found, according to *The Economist* view, allies among agricultural experts and officials and among many deluded consumers and citizens who fall for emotional propaganda about cultural integrity, family values, national security, and the like. All advanced capitalist countries share the problem of self-interested, powerful farmers, in this account, and each can provide an example to represent the others in making one or another point about size of subsidies, strength of lobbies, and so on.

Representing the too small side, radical commentators argue for *Food First*, in the title of a popular and influential work which appeared in the late 1970s, and reflected concerns at the outset of the prolonged crisis of the world food economy. In practice, "food first" was applied to the underdeveloped world and argued for states to favor subsistence production for national markets over commercial production for exports. Like free trade, this is an excellent sounding rhetoric but too absolute to apply realistically to any specific place at any specific time. Several countries, such as Cuba and Nicaragua, attempted to switch from export crops to domestic food crops, with unfortunate consequences, and reversed themselves.

Parallel to the free trader fetishism of markets, food firsters often fetishize specific crops. For instance, potatoes are not widely traded internationally. Does this mean that encouraging peasants to plant potatoes will increase their likelihood of self-provisioning and reduce their dependence on markets for food? No. If a development project succeeds in developing a taste among the local population for a new food, then markets appear, and the most desperate cultivators sell crops for cash rather than keep them for nourishment. Indeed, even longstanding self-provisioning crops, such as tapioca in Thailand and elsewhere, have in the past decade become traded internationally for animal feeds, and self-provisioning farmers have been displaced by export monocultures. In a parallel phenomenon in Asia and Latin America, aquaculture for export to rich customers abroad is depriving small fishers and local inhabitants of a good part of their livelihoods and diets.

Perhaps we can come closer to identifying a way that is just right by redefining the issue. Two related changes occurred in the structures of agro-food production (and in the next section, consumption) during the years of the postwar food regime (1947–1973) and its prolonged and still unresolved crisis (since 1973). These have to do with *power* and with *agency*.

Power in the food economy is either strategic, for those who control wealth, or numerical, for individuals who have the potential to act in concert as workers, consumers, or citizens. *Strategic power* in the food economy has shifted from farmers to corporations. It is easy to miss this change because land has, for the most part, remained the property of farmers and because farmers apparently retain their historic weight in influencing international economic policy. Of course, farms have grown in scale, especially the commercially successful ones (often now legally incorporated), and the gap has widened between large, commercial farms and small, marginal farms (seen in any country or on a global scale). Nonetheless, farmers exist on a human scale compared to corporations. Giant agro-food firms provide inputs (such as equipment, chemicals, seeds, and feedstuffs), and buy the products of farms (such as grain for animal feeds, all the ingredients of manufactured foods and meals, and fruits and vegetables for processing or long-distance shipping).

With the rise of agro-food corporations after World War II, the sources of profit in the food economy shifted from trade to manufacturing, distribution, and food services. Profits in farming had never been an important part either of farm income (which combines returns to family labor, capital, and land), or of national income. The main contribution of commercial agriculture to national economies, both exporters and importers, had been in the steady supply of cheap food which allowed wages to support higher levels of consumption without infringing on profits. Despite their own shifting fortunes, family farmers have consistently subsidized workers and capitalists in the rest of the economy (Friedmann 1978; Warnock 1987; Strange 1988; Bennett 1987; Kneen 1989).

The increasing commodification of food involved the industrialization of agriculture and its subordination to industries supplying inputs and buying outputs. Industrialization of agriculture provided demand for intermediate goods from major postwar industries, especially vehicles and chemicals. It also definitely disrupted the primordial connection between, on one side, the land and labor that cultivates plants and rears animals and, on the other side, the human activities that acquire, transform, prepare, serve, and share meals in daily life and on festive and ritual occasions. Instead of selling products to be consumed after minimal transformation, farmers became suppliers of raw materials to industries producing food which approximated, as much as the laboratories of General Foods could achieve, the status of durable consumer goods (Friedmann 1990).

Paradoxically, national regulation based on farm interests provided fertile ground both for the emergence of transnational agro-food corporations, and for the new importance to agro-food profits of high levels of consumption of complex food commodities. In the emerging agro-food sector, farmers lost their strategic place in production and much of their political clout, while national states eventually found their regulatory capacities diminished through the private restructuring of an international agro-food sector by transnational corporations (Friedmann 1993).

The rise of agro-food corporations and the mass consumption of standard manufactured edible commodities created new *agents* of cooperation and conflict in the agro-food sector. As strategic power shifted from farmers to corporations, *numerical power* in the food economy shifted from farmers to consumers and to food workers in manufacturing and services. This power is always potential and provisional. It depends on the organization of significant numbers into interest groups, such as consumers associations or unions.

Workers in manufacturing and services are far more numerous in the food economy than farm workers of all kinds. While farm workers were always divided between self-employed farmers or (in some unusual places, such as California) corporate employers on one side and migratory, disempowered farm workers on the other (Thomas 1985), the "farm lobby" has overwhelmingly represented self-employed, native-born male farmers. The shift in employment, first in food manufacturing and later in food services—restaurants and fast-food outlets—has shifted the gender and ethnic/race composition of workers in the agro-food sector as a whole.

The shift to services was also part of a larger restructuring of labor forces in advanced capitalist economies, from full-time employment of native-born men, often highly unionized, to part-time and casual employment of women, young people, and ethnic minority workers of all genders and ages. The fast-food outlet and convenience store is a typical site of the growing service sector employment which began in the 1970s. Then the restructuring turned back on the old manufacturing industries themselves. Many had always employed women and ethnic minorities and had resisted unionization, especially in the food-processing industries. However, even in the very old, highly unionized industries, such as meatpacking, corporations in the 1980s reorganized, relocated, and replaced native-born, unionized male workers with unorganized female and ethnic minority workers (Ufkes 1991; Stanley 1994). Struggles by the Meatpackers Union resisting relocation and downgrading of wages and working conditions were widely reported in the national media. Even within farming itself, women have begun to assert their visibility and importance, previously hidden under the rubric of the "farmer's wife."

At the same time, with the rise of mass consumption as a key to economic stability, consumers became significant to the profits of agro-food corporations. The people who concern themselves most consistently with consumption (in economic terms) or the provisioning of their families (in social terms) are overwhelmingly women. This has always been true in some senses, but the restructuring of daily life in the postwar period created new roles for women. Suburban development and the miniaturization of domestic appliances for use in shrunken nuclear family (and now even smaller) households created the spatial and technical conditions of work for the postwar American housewife.

Separated socially from folk knowledge of household organization, child rearing, and food preparation and separated physically from markets as sources of daily provisions and sociability, women who adopted the new postwar norms of domesticity became crucial to the larger economy both as "shoppers"— which became a specialized, skilled new task (Glazer 1990)—and as "managers" of houses equipped with the increasing panoply of consumer durables (cars, televisions, washers, dryers, and smaller appliances, including refrigerators, stoves, freezers, and others related to food storage and preparation)—which kept the large-scale industries of the postwar boom humming. The new foods they bought were often the inventions of corporate "kitchens" rather than culturally transmitted recipes, and they were intimately linked to the use of the new appliances that permitted weekly shopping, long storage, and new ways of preparing meals for families whose gender and age relations were radically restructured (Ewen 1976; Hayden 1984).

As the size of markets became important for agro-food corporations, the poor came to be incorporated as consumers. In the postwar period, various distributional policies have made food (via U.S. food stamps, for instance) or money to buy food (via transfer payments) available to poor consumers. Now, after four decades of steady growth, consumption in the 1990s threatens to shrink because poor people and debt-strapped countries need more subsidies, which are not forthcoming. As the distribution of income becomes more unequal, privileged consumers are decreasing as a proportion of total demand. Their consumption can produce only limited profits no matter how refined their tastes become. As poverty becomes feminized, and ethnically concentrated, the politics of distribution coincide with the politics of employment.

The gender and ethnic character of food politics is bound to change along with changes in consumption and production. Therefore, despite the high media profile of farmers, and their magical empowerment by molders of public opinion, such as *The Economist*, new agents are potentially more important to the future of agro-food politics. First are paid workers in the food sector. These are increasingly women, young people, and ethnic minorities, usually without the stability, wages, or collective organizations that typified the white, male, unionized workforce in the heavy manufacturing industries (including meatpacking) during the decades of the postwar boom. Second are unpaid workers in the domestic sphere. These are overwhelmingly women burdened with the increasingly complex tasks of finding money to feed their families and negotiating the welter of edible commodities in increasingly global marketplaces.

People trying to find or construct stable, socially useful work, physically and socially satisfying diets, and habitable communities all have issues distinct from those now dominating formal food politics. Economists and corporate managers, who have considerable clout in setting political agendas, count the human costs of hunger and the ecological costs of monocultural farming as "external." Therefore, the problems that get posed in the political arena concern distribution (size and location of markets), location of food manufacturing and services, and conditions of employment. Environment, health, land use, poverty, and a variety

Food Politics • 25

of issues in principle related to *food policy* for the most part remain separate policy issues, subject to separate administrative agencies. Agricultural policy is at an impasse because it cannot address these social problems. New agents can in principle find unity through redefinition of issues centered on the production and consumption of food.

Before turning to politics and policy, however, it is important to look at how consumption has changed, and might change. As food products became more important as a source of profits, so did the practices of consumers and the attempts to influence them. Changes in diets, therefore, are crucial to the future of agro-food power, profits, and livelihoods. The degree of choice available to individuals varies by region, nation, and class, and by age, gender, and ethnic group. The ways that individuals exercise the choices they do have, or struggle to have more choices over their diets, is the next part of the story.

CONSUMPTION: POWER AND DESIRE

Like Goldilocks, too, in the sphere of consumption, we are offered explanations for what people eat which are either too soft or too hard. Either, according to neoclassical economists, individuals are completely free to choose foods in the market, according to their purchasing powers and desires, or, according to food radicals, their minds are "colonized" through advertising and merchandising which plays on the desire of human beings "to imitate those they perceive as their social superiors" (George 1984, 89–90).

Why do some people change diets and some not? Why do some categories of people—nationalities or classes—change and not others, and at some times and not at others? More important, how does the influence of corporations, advertisements, and so on, which is often attributed to capital, actually work, and when does it fail? Why do some people stick with or give up curry? In contrast to a lot of literature on the spread and dominance of the American or corporate diet, I find it easier to see why people keep their diets, or adapt them to new settings and conditions, or do syncretic things with them, than to see why they behave as if they actually prefer Coke and hamburgers to the foods associated with their memories and social lives, especially when these new dishes often cost so much money.

Change of diets is never given in advance, but results from a provisional resolution of various tensions involving possibilities and limits, fears, and desires. These are negotiated by individuals in situations with varying balances between stability and change. For example, while MacDonald's is opening new outlets throughout Asia, at least some Asian immigrants to Toronto are retaining their diets and making the necessary adaptations to do so. Recently, as I waited for a tuna sandwich in my college cafeteria, I was entranced by the unexpected scent of curry. It came from a microwave oven being used by a student to warm a dish brought from home in a microwavable container. It was, at least at the point of consumption, more convenient as well as (I thought) more desirable than the convenience foods offered to me commercially. Yet did this student's resistance to commercial, standard food right in the heart of North America rest on her own or her mother's unpaid labor of shopping and preparation? Who can or wants to do that, and who doesn't? Is that relation of apparent choice between (one's own or someone else's) money and time, as well as taste, the same in Toronto as in Lahore?

The explanation that is just right, I think, must link the *power of desire* to relations and practices that express *social power*. Food and diet, because of the material importance and emotional charge they carry, are useful windows on power as a complex field of social relations and symbolic representations. It posits the necessary presence of *both* objective possibilities *and* subjective desire. This encourages us to specify the concrete relations and objects, opportunities and constraints, in each situation. For just as cultivators accept or resist specific plants, animals, and the practices associated with them, so people accept or resist specific ingredients, modes of preparation, and meals. To understand the mix of acceptance and resistance in each time, place, and culture, then, we must include both what people do with their actual choices and what they think and feel about what they do (Mintz, ch. 1, this volume).

The power of desire and social power are linked through the *social relations of consumption*, which are based in the daily life experiences of preparing, sharing, and taking meals. The social relations of consumption is a concept parallel to the social relations of production. Both were radically reconstructed in advanced capitalist economies, particularly in the United States, in the postwar years. Without the new social relations of consumption, created through the disruption of old patterns of family and community, it is doubtful that so many people would have come to desire the new foods, such as Miracle Whip, Velveeta, TV dinners, and the like, whose mass demand was crucial for the profits of such emerging food manufacturers as Kraft and General Foods, in addition to those of the service franchises whose fried chicken and hamburgers set new standards in place of each family's recipes—just as Dr. Spock and women's magazines set common standards in place of folk wisdom about child rearing and running a household. (These comparisons make clear, I hope, that improvements may sometimes have resulted, both in respect for children and standards of hygiene. The point is that expert knowledge replaced folk knowledge, in nutrition and recipes, as in other aspects of domestic life.)

In my view, it makes sense to interpret changes in diets and cuisines (Weismantel 1988, 87) in terms of a shifting balance between coercion and consent, both in the constitution and reconstitution of desire for specific foods, and in the availability of those foods. When the balance is heavily weighted toward coercion, we can think of a *despotic dietary regime*. When the balance is weighted toward the other extreme of consent, we can think of a *hegemonic dietary regime*. This is an adaptation to the sphere of consumption of Michael Burawoy's (1985) distinction between despotic and hegemonic regimes of labor control.

The question of the balance between coercion and consent in relations of consumption arises when people's lives are disrupted so that they cannot continue to use land, to work, and in other ways to live as they did before. This often occurs, as Mintz (ch. 1) argues, by violence and war. It happens more persistently and more subtly through changes in entitlements (Sen 1982) in the wake of commodification—of land, of labor, and of food. This means first that the link between food and money changes both the availability and the meaning of specific foods. Kate Soper (1981) has argued that human needs become shaped by general commodification, so that the basic human need becomes the need for money. Physiological needs for nourishment, psychological needs for nurturance, and social needs for commensality, ritual, display, celebration, and much else are all shaped by the varying ratios at which specific foods exchange for money.

In a fully commodified society, money expresses social power. For those who have plenty, money commands the produce of the world. For those who lack money, desire is perforce shaped by the relative costs of acquiring and preparing meals. This difference inspires Soper's (1981, 63) criticism of the frequently heard, earnest assertion that basic needs, such as food, are actually simple. She asks rhetorically whether the smoked salmon and out-of-season asparagus eaten by a corporate manager are actually supplying the same need for food as the reconstituted Chinese dinner eaten by his worker. It is clear, she argues, that a 300-calorie meal can contain twenty times the social wealth of a 6,000-calorie meal. This fact is not lost on the manufacturers of "diet foods," the most paradoxical source of profits in a hungry world.

In partly commodified societies, where what people cultivate is intimately bound up with what they eat, the effects of money can be less visible but more pernicious. People without enough food must make dangerous choices to survive, choices deeply affected by the relative prices of foods in the market. An example comes from a self-provisioning area of Mexico, where the combination of maize, beans, and squash is rooted in ancient myths, long-standing horticulture, and richly elaborated cuisines. With population pressures and commodification of land and labor, people often respond to the shortage of calories by selling squash at relatively high prices to supplement their inadequate home production of maize. This works as a short-term strategy, but in the long term, it results in a vitamin A deficiency and gradual blindness (Dewey 1981).

At the same time, the meaning of specific foods changes with social relations of consumption. Once again, the lowly spud is a fine example. In contrast to the strenuous resistance to the potato in eighteenth- and nineteenth-century Europe, the Irish found it the only crop that allowed them to survive centuries of English colonization, land appropriation, mercantile restrictions, exploitation, and religious persecution. They did so by reverting to self-provisioning, at the same time retaining their practice of partible inheritance. Over time, the self-provisioning plots became ever smaller and gradually their dependence on potatoes became nearly complete. Eventually the potato crop succumbed to blight.

Subsequently, both in Ireland and in the lands of Irish emigration, this survival food became the object of desire. In the desperate aftermath of flight from starvation, the potato changed from a means of physical survival to a means of cultural survival. It remains a defining ingredient of the cuisine of the descendants of the survivors of the nineteenth-century Irish famines.

As we saw, partly because of its association with the Irish, the potato was resisted by English workers as an attempt to lower their standard of diet by replacing the more expensive wheat. The contest was eventually won by employers and statesmen. Within a few decades, the potato had acquired the status of tradition in meals of meat and potatoes, fish and chips. Similar contests had similar results elsewhere, reflected in all the potato meals and drinks of Northern Europe. Yet the contest between ordinary people and those who would impose changes in their diets never ends. It is renewed over new foods when circumstances warrant, either for those seeking to acquire the foods of the privileged, or for those seeking to manipulate the desires of the poor.

Even the struggle over the potato had a provisional outcome, expressed in its ambivalent role as both a lowly staple and a traditional food. When social reformers began to codify a hierarchy of food preferences as part of the anatomy of poverty, the potato was at the bottom. The standard of living was thus measured in part by the rapidity with which people switched to higher foods, such as meat and fresh fruits and vegetables, as their incomes increased. That was the 1930s. Beginning in the 1950s, consumption of frozen potatoes increased astronomically. Although they remain less than half the per capita weight of fresh potatoes in U.S. consumption, frozen potatoes represent many multiples of the value of fresh potatoes. A bag of chips takes further the shift in value between the humble root and all the machines, hands, oils, chemicals, packaging, and advertising it has passed through.

So what can we make of the soft and hard approaches to how and why people change their diets? The balance between consent and coercion is variable and provisional, but it is always present. Therefore, "consumer sovereignty" and the "colonized mind" are positive and negative versions of the same fantasy: a world of imaginary people with no history and no social ties, either by happy individuality or by unhappy susceptibility to manipulation. The just right, or balanced approach, has the merit of not giving us answers in advance as to whether or not people will change their diets.

George's point about imitation holds up, I think, only if we see it in the context of social relations that create differing balances between coercion and consent. The desire for abundance always fixes on concrete images, whether bags of gold or lusciously detailed descriptions of meals consumed by kings, princesses, movie stars, or bond brokers. Advertisements do in pictures what fairy tales and myths have always done in words. Although we know that abundance does not entail satisfaction, any more than beauty ensures a prince charming, still deprivation probably intensifies susceptibility to manipulation by images. One of the first advertisements made after World War II for a global audience promoted Pepsi-Cola. It introduced a culturally alien and nutritionally dubious product into any telecommunicatory nook and cranny the marketers could find. Made during the McCarthy era, it was wordless and depicted happy, healthy, tanned, young Americans playing on beaches, in sports events, and other pleasure sites, all swigging Pepsi bottles between smiles. Anyone working hard, not to say wondering where food for the family would come from, would have reason to imagine himself or herself into that picture.

Modern advertisements offer, like the mythical palace granted by the genie to the lucky fisherman, concrete images of paradise. They depart from fairy tales by suggesting that the objects of desire are universally accessible, and that something other than luck, something each person can do, will bring them closer to paradise.

ALTERNATIVES

What food politics are just right? Global capital is dancing as fast as it can. Land and labor are increasingly disembedded from the social relations in which food is produced and consumed. More and more goods are sold to global markets, and more and more goods are purchased from those same global markets. Not only are land and labor being more intensively turned into fictitious commodities, but money is increasingly coming to mediate all food relations. Through pressure to export and reduce import barriers, the principle of distance, and its associated principle of durability, are fast becoming dominant in an increasingly global agro-food sector.

The consequences are plain. Insecurity is growing in rich and poor countries, and more and more people are falling, or fear that they may fall, from the ranks of privilege. More people are hungry than ever, and more employment than ever, in one's own right or for wages, is tenuous. Environmental degradation is growing, and its consequences are increasingly recognized to be global. Organic, diverse farming is being destroyed more quickly than technical solutions are found to the problems created by chemical-intensive monocultures.

Yet state regulation of agriculture is no longer appropriate to the problems of hunger, employment, or sustainable land use. Farmers are the most vigorous defenders of present regulation, out of self-interest, to be sure. But they have considerable public support, in Europe especially, because the alternatives are so far posed only in the negative: deregulation (in favor of corporate regulation?); "decoupling" of agricultural prices from farm subsidies (in favor of links between prices and corporate buyers? between farm subsidies and what other state expenditures?); removal of protection for local products, such as wines, cheeses, and prepared foods of all kinds (in favor of standard products of the global corporate laboratories, bearing descriptions as "French," "Thai," or "Mexican"?).

The real alternative is democratic regulation of regional food economies. Let me take each element in turn. *Food*, not agriculture, is the appropriate unifying principle of politics and policy. The problems associated with food as a social need have shifted since the 1930s, when present forms of regulation were devised. The level of production is now less important than the *location, diversity,* and *scale* of farming. Agriculture has, in fact, been subordinated to food industries and services by agro-food corporations during the past half-century. To build on that legacy creatively, communities can redefine agricultural policy within the framework of *food policy*, which links what is produced, how, and by whom to socially desirable employment, just access to quality diets, and sustainable, socially beneficial land uses.

Only food economies that are bounded, that is, *regional*, can be regulated. To create regional food economies requires politics that re-embed land and labor in the needs and capacities of communities. Land and labor are currently "productive" only when required by employers, which are overwhelmingly corporate and increasingly global. Yet corporate principles of distance and durability can in some ways be bypassed. Practices that relink what people can do with what we need promise to re-embed the economy in social life. Of course, people all over the world seek their livelihoods creatively. They barter, exchange, and find innovative ways to protect themselves and each other (as well as take advantage of each other).

These strategies are often feeble in comparison to the power of corporations and the "market," yet it is one area in which First World citizens have much to learn from strategies found throughout the Third World. There livelihoods have longer been precarious and the experiences of life before commodification are wider and more recent. Community kitchens, direct links between urbanites and farmers, including investment and guaranteed sale, including informal groups and larger cooperative ventures, are examples of links that can be forged in the shadow of the self-regulating market. These are more significant when they also favor agronomic practices that enhance the soil, water, and other features of the environment.

If food is to be susceptible to democratic regulation, the links in the food chain must first be made visible. An environmentally and socially sensitive agriculture presupposes consumers whose food needs are effectively transmitted to farmers, as well as citizens whose environmental needs are effectively transmitted to farmers. This can happen through markets and other institutions consciously regulated by governments at every scale, beginning with the smallest (Sale 1980; Dauncy 1986). Whatever needs cannot be met at the smallest scale, corresponding as much as possible to the natural features defining bioregions as well as the legacy of social settlements, can be regulated at higher, overlapping levels. This focus on the smallest scale, however, can work only if it is supported by higher levels.

At present, national controls are the strongest in existence, although they are growing weaker through the privatization of regulation by corporations from above, and the movements for autonomy from below. It is important neither to defend present regulation, which is ineffective, nor passively to support the corporate agenda simply to deregulate. Active politics to transform regulation as coherent *food policy* can link concerns about employment, access to food, quality of food, and sustainable agriculture.

At the same time, however difficult, it is essential to create politics across nations to create democratic public regulation at the international level. Just as regional politics (which might be municipal, state/provincial, or other) depend on supportive national institutions, so national regulation depends on supportive international institutions. The present agenda at international fora, such as the General Agreement on Tariffs and Trade (GATT), or the North American Free Trade Agreement, is to dismantle formal regulations but not regulatory institutions. Indeed, international regulatory institutions are being strengthened—note the enhanced powers of the IMF or the proposed new powers of the GATT to police ''intellectual property rights'' in supposedly sovereign national jurisdictions. This is paradoxical only if we fail to see that public regulation is being replaced by private, corporate regulation to be enforced by newly empowered international institutions.

Workers and consumers in the food system can best defend and enhance their welfare by understanding that agriculture must become more labor intensive to be sustainable, in most cases, and that food processing and distribution will create local employment if the links are kept as close to the region (in nests of regions of increasing size) as possible. Governments can change taxation and credit policies, which currently favor chemical-intensive monocultures, to encourage better agronomic practices and local networks of farmers and local enterprises that process and serve foods. Governments can use public buying power, for school lunch programs, hospitals, offices, and the like, to increase the markets for local products and set standards of food quality determined by political participation. An economy embedded in the community not only serves the individuals who benefit from employment, but also reflects the cultural diversity of its citizens and allows for experimentation and change.

This is by no means to advocate nostalgia or a return to past practices, which entail past scarcities and forms of oppression. If science and technology were to serve regional food economies, much is possible. Some advances are already in existence, such as glass houses and vacuum packing, which increase the range of local products. Others, such as the potential use of field crops as renewable energy sources, suggest the range of possibilities and the institutional changes necessary at national and international levels. If we were to subject the potential of biological innovations to democratic plans for food and health, some of the visions of dynamic and sustainable economies might be closer to realization.

Corporations, agro-food workers, and consumers are the new contestants over food rules. Their potential allies are government agencies at various levels, from municipal to international; agriculturally marginal groups, such as ecological farmers; and social movements concerned with income and food security, with environmental sensitivity, and with health. The balance of power held by agro-food corporations can shift only if workers and consumers ally with environmentalists and social justice movements, as *citizens*, to create and institute democratic and appropriately scaled food policies.

REFERENCES

Bennett, John. 1987. *The Hunger Machine*. Montreal, Quebec: CBC Enterprises.

Burawoy, Michael. 1985. *The Politics of Production*. London: Verso.

Dauncy, Guy. 1986. "A New Local Economic Order." In *The Living Economy: A New Economics in the Making*, edited by Paul Ekins, 264–72. London: Routledge & Kegan Paul.

Dewey, Katherine. 1981. "Nutritional Consequences of the Transformation from Subsistence to Commercial Agriculture in Tabasco, Mexico." *Human Ecology* 9, 2: 151–87.

Economist. 1992. "The New Corn Laws." December 12.

Ewen, Stuart. 1976. *Captains of Consciousness: Advertising and the Roots of the Consumer Culture*. New York: McGraw-Hill.

Friedmann, Harriet. 1978. "World Market, State, and Family Farm: Social Bases of Household Production in the Era of Wage Labor." *Comparative Studies in Society and History* 20: 4, 545–86.

———. 1990. "Family Wheat Farms and Third World Diets: A Paradoxical Relationship between Unwaged and Waged Labor." In *Work without Wages: Domestic Labor and Self-Employment within Capitalism*, edited by Jane L. Collins and Marta Giminez, 193–213. Albany: State University of New York Press.

———. 1993. "The Political Economy: A Global Crisis." *New Left Review* 197: 29–57. Also forthcoming in *Food*, edited by Barbara Harriss-White. Oxford: Blackwell.

George, Susan. 1984. *Ill Fares the Land: Essays on Food, Hunger, and Power*. Washington, D.C.: Institute for Policy Studies.

Glazer, Nona. 1990. "Servants to Capital: Unpaid Domestic Labor and Paid Work." In *Work without Wages: Domestic Labor and Self-Employment within Capitalism*, edited by Jane L. Collins and Marta Giminez, 142–67. Albany: State University of New York Press.

Goodman, David, Bernardo Sorj, and John Wilkinson. 1987. *From Farming to Biotechnology*. Oxford, England: Basil Blackwell.

Goody, Jack. 1982. *Cooking, Cuisine and Class*. Cambridge, England: Cambridge University Press.

Hayden, Dolores. 1984. *Redesigning the American Dream: The Future of Housing, Work, and Family Life*. New York: W. W. Norton.

Kneen, Brewster. 1989. *From Land to Mouth: Understanding the Food System*. Toronto, Ontario: New Canada Press.

Lappé, Frances Moore, and Joseph Collins. 1974. *Food First: Beyond the Myth of Scarcity*. Boston: Houghton Mifflin.

Mintz, Sidney. 1979. "Time, Sugar, and Sweetness." *Marxist Perspectives* 2: 56–73.

Polanyi, Karl. 1956 [1944]. *The Great Transformation*. Boston: Beacon.

Salaman, Redcliffe, 1985. *The History and Social Influence of the Potato*. Cambridge, England: Cambridge University Press.

Sale, Kirkpatrick. 1980. *Human Scale*. London: Secker & Warburg.

Sen, Amartya. 1982. *Poverty and Famines: An Essay on Entitlement and Deprivation.* Oxford, England: Clarendon.

Soper, Kate. 1981. *On Human Needs: Open and Closed Theories in Marxist Perspective.* Atlantic Highlands, N.J.: Humanities Press.

Stanley, Kathleen. 1994. "Industrial Change and the Transformation of Rural Labor Markets in the U.S. Meatpacking Industry." In *The Global Restructuring of Agro-Food Systems*, edited by Philip McMichael. Ithaca, N.Y.: Cornell University Press.

Strange, Marty. 1988. *Family Farming: A New Economic Vision.* Lincoln: University of Nebraska Press.

Super, John C. 1985. "The Formation of Nutritional Regimes in Latin America." In *Food, Politics, and Society in Latin America*, edited by John C. Super and Thomas C. Wright. Lincoln: University of Nebraska Press.

Thomas, Robert J. 1985. *Citizenship, Gender, and Work: Social Organization of Industrial Agriculture.* Berkeley: University of California Press.

Ufkes, Frances. 1991. "The Changing Social Structures of Livestock Marketing in Illinois, 1950–1990." Paper presented at the Association of American Geographers, Miami, Florida, April.

Warnock, John. 1987. *The Politics of Hunger.* Toronto, Ontario: Methuen.

Weismantel, Mary J. 1988. *Food, Gender, and Poverty in the Ecuadorian Andes.* Philadelphia: University of Pennsylvania Press.

3

Durable Food Production and Consumption in the World-Economy

Susan J. Thompson and J. Tadlock Cowan

From the mid-sixteenth century, the region from Cape Cod to Labrador was the center of one of the world's richest fishing grounds (Wallerstein 1974). This region was also the center of one of the New World's first agro-food industries: the production of dried and salted cod. Salted and dried cod, along with the salted herring of the Baltic area, were among the first foods made "durable" by simple preservation techniques at the site of harvest. These commodities were then capable of being transported to the metropole where they became an important component of urban diets. During the seventeenth century, the North Atlantic region also became a provisioning source for the plantation economies of the western Atlantic periphery: the southern United States, the West Indies, and Brazil. Although the market for salted and dried cod from the North Atlantic region declined in the eighteenth century (Ommer 1991), this region reasserted itself in the global economy in the nineteenth century with a new preserved fish—tinned herrings or sardines.

In this chapter, we examine the role "durable foods" have played both in the development of a global agro-food system and, with that development, in the expansion of the world-economy. In particular, we discuss tinning or canning as a technological and commodity food link in the development of the first food regime and in the further elaboration of high value-added, manufactured foods characteristic of the second food regime.[1] Our particular focus is seafood, an important animal food in the metropolitan diet since the Middle Ages, that also provided a cheap source of protein to the urban poor. Seafood is a commodity that has been preserved and traded since the Bronze Age. It is, moreover, one of the first foods to be canned and one of the first to enter world trade.

The invention in 1810 of iron containers plated with tin to protect against corrosion, giving further durability to a highly perishable commodity, permitted the large-scale development of this form of food preservation. Canned seafood remained an affordable food and consequently played an important role in feeding the military and, later, industrial workers. We discuss canned seafood as an important early industrial food that predates the more intensive development of food manufacturing the post–World War II period. Finally, we discuss the development of canned seafood as a global commodity introducing metropolitan diets into the periphery.

CONCEPTUALIZING DURABLE FOODS IN HISTORICAL CONTEXT

Research on the historical development of agro-food systems in the world-economy has emphasized how changes in capitalist accumulation have deepened commodity relations in food production and consumption (Friedmann 1991; Friedmann and McMichael 1989; Kenney et al. 1991). This research, focusing on changes occurring after World War II, has discussed the importance of durability in the development of agro-food systems, in changing forms of food commodities and their consumption, and in the role of nation-states and transnational corporations in organizing these changes. Durable food production here includes the transformation of agricultural commodities and foods to either intermediate raw materials or highly processed foods.

While a critical link in the intensification of food production and consumption relations during the second food regime, durable foods predate industrial food processing and the first food regime. The term "industrial food" has been used in reference to the diets of the working urban populations of nineteenth and early twentieth century Britain as well as to the technologies of the period involved in the preservation of food (Goody 1982). In this chapter, the term industrial foods is used to identify the distinctive production technologies and relations of consumption that prefigure the second food regime as well as better to locate historically the changes in food durability technology that provided the bridge to these more complex forms of food production and consumption. We wish to emphasize the relations between the methods of food production and their corresponding relations of consumption. Salting, drying, oil curing, spicing, and smoking, for example, are ancient methods for processing highly perishable foods for trade and for feeding urban populations (Garnsey 1988; Reid 1988; Tannahill 1973). With the development of refrigeration technology in the late nineteenth century, freezing also became an important processing method for the world food trade. Prior to that innovation, however, canning or tinning was the dominant industrial food-processing method. Canning or tinning provided a technological linkage between the ancient methods of food preservation and the more modern ones, such as freezing, condensing, and freeze-drying, and the new manufactured foods and thus played an important role in structuring agro-food linkages in the world economy. Tinned foods were also important to an expanding industrial workforce, in part because they did not require the added expense of further technology or labor to maintain them or prepare them for consumption (Tannahill 1973; Renner 1944). Freezing, in many ways a superior method of preservation, nonetheless required the food purchaser to consume the food quickly or to store it in some household or community cold-storage unit for later consumption.

We believe that the important concept of durable foods can be further developed by linking it to earlier processes of food preservation, the latter being, in our judgment, the more historically generalized phenomenon. Although Friedmann (1991), for example, clearly recognizes the role that technology has played in extending the shelf life of foodstuffs (e.g., through refrigeration), her analyses have not detailed the historical role played by durable foods in the formation and expansion of food consumption patterns in the world system. Here we take some initial steps at identifying the linkages between various stages and technologies of durability, conceptualizing durable foods as a *continuum* that includes preserved foods characteristic of preindustrial diets (e.g., salted cod, dried meats and fruits, and wine) and the manufactured foods characteristic of industrial diets in the second food regime (Friedmann and McMichael 1989; Braudel 1979; Friedmann 1993). Tinned or canned foods thus represent an important technological and commodity link between simply preserved foods for direct consumption and the high value-added manufactured foods characteristic of the current agro-food system (Friedmann 1993).

Durable foods refer, very generally, to those foods still consumable after an extended period. Goody (1982), in an important earlier work, similarly calls these foods "long-life" foods. Durable is, of course, a somewhat imprecise term. The transformation of milk into yogurt, cheese, and butter, for example, gave longer life to milk, but, in the process, created new foods that, while extending the usability of milk, were nonetheless limited by the rapid perishability of dairy products. In addition, these milk "extenders" still required some form of cool storage. Canning, on the other hand, has been known to preserve foods, including fluid milk (e.g., condensed and evaporated), for years without altering the basic form of the food, thus making an industrial product that was closer to the fresh food commodity. When considering a food's durability, then, it is useful to think of foods existing along a continuum from those that are simply *preserved* (e.g., olives and dried fruits), to *processed* (e.g., frozen "french" fries and tomato paste), to wholly *manufactured* foods (e.g., Jello, instant coffee, Tang, most breakfast cereals). We define processed foods as those whose initial foodstuffs have been altered before they are marketed for final consumption. Manufactured foods use foodstuffs as intermediate commodities in the production of final, high value-added commodities (Friedmann 1993).

The food durability continuum is associated with the parallel development of a range of preservation technologies. This food durability continuum, thus, is one dimension of a more complex model (Figure 3.1). Food durability can be considered the horizontal axis of a two-dimensional model of food preservation whose vertical axis is food preservation technology. Prior to the first food re-

Figure 3.1
Durable Foods

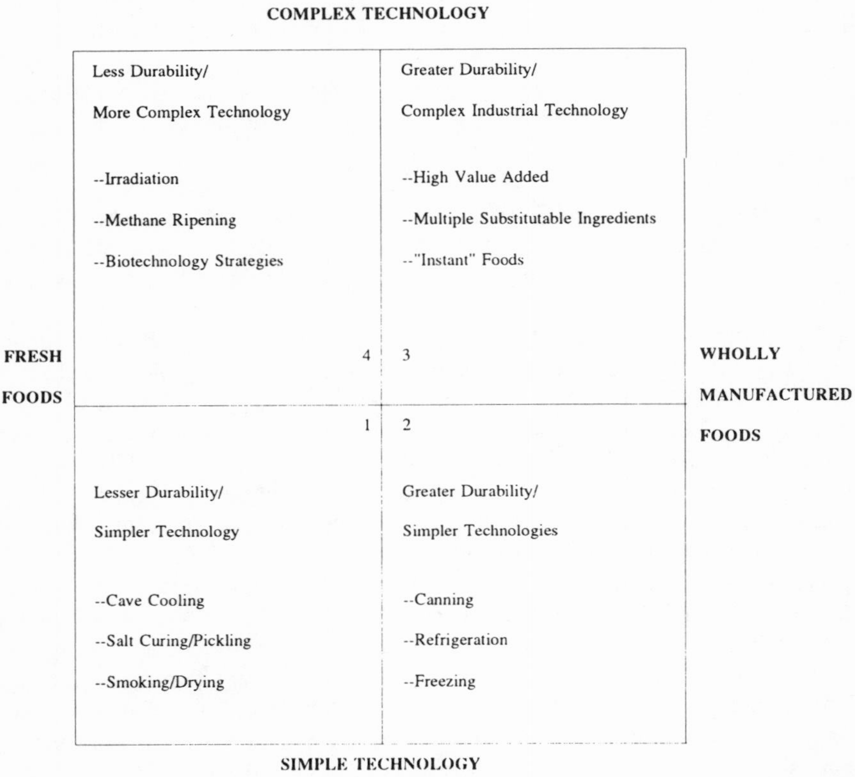

COMPLEX TECHNOLOGY

Less Durability/	Greater Durability/
More Complex Technology	Complex Industrial Technology
--Irradiation	--High Value Added
--Methane Ripening	--Multiple Substitutable Ingredients
--Biotechnology Strategies	--"Instant" Foods

FRESH FOODS   4   3   **WHOLLY MANUFACTURED FOODS**

   1   2

Lesser Durability/	Greater Durability/
Simpler Technology	Simpler Technologies
--Cave Cooling	--Canning
--Salt Curing/Pickling	--Refrigeration
--Smoking/Drying	--Freezing

SIMPLE TECHNOLOGY

gime, the food durability continuum ranged from fresh foods with a fairly long storage life (e.g., root vegetables) to those very simply preserved. The preservation technology continuum ranged from storage techniques, such as root cellars, to the simple preservation technologies of drying and pickling (quadrant 1 in Figure 3.1). In the first food regime, the food durability continuum extended beyond simple preserved to simple processed foods (e.g., jellies, ketchup, mustard, canned corned beef). The preservation technology continuum ranged from sun drying and smoking to tinning and early refrigeration and freezing (quadrant 2 in Figure 3.1).

The second food regime's food durability continuum is somewhat similar, ranging from highly preserved foods (e.g., frozen vegetables) to more complex manufactured foods (e.g., multiple substitutable ingredients combined into new food commodities) for increased mass consumption by urban elites in the periphery (quadrant 3 in Figure 3.1). Examples include instant coffee in major coffee-producing areas and instant breakfast drinks in major citrus-producing export areas. It is the extension of the preservation technology continuum that significantly expanded food durability, particularly in the postwar era of mass consumption. The durability axis can, in the emergent food regime also extend to fresh food as well as to wholly manufactured foods. The preservation axis here extends from the simplest and oldest preservation technologies, such as sun drying, to emergent food technologies based on biotechnology, gas ripening, and irradiation (quadrant 4 in Figure 3.1). The focus here is on "designing" durability into the raw food commodity itself. Commonly, fresh foods are those that can be picked, plucked, gathered, and consumed without subjecting them to any storage process other than the simplest forms of preservation, for example, cooling. Gamma irradiation and biotechnology are currently targeted at vegetables and fruits because there is high demand (Bennett's Law) in core countries.

DURABLE FOODS IN AN EMERGING GLOBAL ECONOMY

Food preservation is as old as human civilization. Preservation technologies transform perishable animal and vegetable matter into edible, stable, and portable (and thus durable) commodities. Perhaps most important of all, preserving makes possible the transport and long-term storage of food products. Preservation of foods inhibits spoilage caused by bacterial growth, oxidation, insects, or desiccation. Fermentation, oil packing, pickling, salting, and smoking are all ancient preservation technologies. Milling and baking, brewing, and cheese and yogurt making are ancient methods of food processing that also extend the durability of foods. Refrigeration in caves or under cool water were also well-known ancient techniques of food preservation. Most other methods of preservation are inventions of the past 200 years. Mechanical refrigeration, various dehydration techniques, and quick freezing were late nineteenth- and early twentieth-century discoveries.

The technical ability to preserve foods was an important factor in the expansion of the world-system through the growth of trade. Prior to the nineteenth century, drying and the application of salt, spices, or sugar were the dominant methods of food preservation. Salting, which was probably the most important preservation technique, was central to provisioning an expanding urban population. Changes in salting technologies at the end of the fifteenth century enabled the provisioning of sailors and urban wage labor for the next four centuries (Braudel 1979; Rediker 1987; Drummond and Wilbraham 1939). For sailors in the Atlantic, salted meat was a major part of the seaman's weekly diet of "one pound of meat—salt beef (called 'Irish horse' [because most salted meat was produced in Ireland]), pork, bacon, or at times fish—four times a week, with additional portions of cheese, peas, butter, and biscuit, and rations of wine, brandy, small beer, or rum" (Rediker 1987, 127; Braudel 1979); in the Mediterranean region, salted fish remained the standard fare for sailors (Braudel 1979).

Salt preservation technologies were long and widely known and salted herring was an important staple of urban and military diets from the Middle Ages (Jenkins 1927). The designation of certain religious days as "fish days" by the Catholic Church further increased the demand for fish. This demand for fresh fish, however, could not be met from local sources alone, and, by the middle of the sixteenth century, fish days had become an important stimulus to the shipbuilding trade (Drummond and Wilbraham 1939). Until the establishment of turnip husbandry permitted winter feeding, cattle and other meat animals were slaughtered in the fall and most of the meat was consumed during the early winter months, hence the importance of fish and Church fish days, not least for the pacification of hungry urban populations. Between the sixteenth and nineteenth centuries, salted foods continued to grow in importance not only in urban diets but also in the diets of the plantation laborers in the Americas (Braudel 1979; Smith 1990; Ommer 1991; Wallerstein 1980).

During this period, however, quality was very difficult to maintain, and great variation existed in the durability of salted fish. The state's interest in the quality of preserved foods also emerged at this time. The Baltic and (later) North Sea salted herring market, for example, was well regulated to ensure high quality. Quality was difficult to maintain, however, for several reasons. First, the type of herring salted varied. Herring with a high fat content, caught early in the season prior to spawning, were of the highest quality because they absorbed the salt more effectively (Smith 1990). Controlling the fishing season was, consequently, one of the earliest forms of state regulation over food fish. A second factor determining salt fish quality was packing. In order to ensure quality, the size of the fish, the thickness of the layers of barreled fish, and even the type of salt used were stipulated in Dutch regulations. Initially, packing aboard ship was prohibited. Regulations required that salting and barreling herring take place in designated areas ashore. As the location of herring fishing moved to the North Sea and as fishing technology improved, later regulations permitted first-stage packing on the fishing boats, with the fish later repacked under Dutch government supervision in port (Jenkins 1927). These regulations represent early efforts made by the mercantile state to regulate the quality of food consumed by metropolitan populations.

Industrial labor and maritime labor were important groups of workers in the early nineteenth century. Industrial labor was located in densely populated urban areas without access to lands to provide for the subsistence needs of the household (Ellis 1977), while maritime labor needed supplies of food and drink for voyages lasting as long as one year. Each group required provisioning, and the types of foods available were generally similar. The cheapest durable fish foods were, not surprisingly, the poorest quality and were consumed largely by the urban poor and the plantation labor in the periphery. The salted herring of the Shetland Islands, for example, well known for its inferior quality, was excluded by quality control regulations from the German and Russian markets, but it found a ready market in the West Indies and Ireland (Smith 1990). Even cod, as important a food fish as it was, varied greatly in quality due to packing.

Again, low-quality cod was most often found in the diets of the urban poor (Braudel 1979).

Food quality, diet, and class location are strongly related. In urban Britain, the diet of the poorest classes consisted of adulterated bread, partially decomposed meat and fish, and rotted potatoes. As wages increased somewhat, their diets became slightly more diverse, consisting of bread, oatmeal, bacon, a little butter, molasses, tea, and coffee (Drummond and Wilbraham 1939, 394). Skilled workmen had even more meat and a few vegetables other than potatoes. It was estimated that for these "ordinary people," daily consumption was from 3/4 to 1 pound of bread and meat, from 1/4 to 3/4 of a pound of green vegetables (mostly cabbage), from 1/2 to 1 pound of butter, and from 1 pint to 1 quart of beer (Drummond and Wilbraham 1939, 396). But even for these more skilled workers, the meats were the cheapest cuts, although the meat was less likely to be diseased or rotted (Drummond and Wilbraham 1939).

Perhaps even more important than urban demand to the development of durable foods was the need to provision the armies and the expanding merchant and naval fleets of emerging European states. The British Naval Provision Scale of 1785, one of the earliest records documenting sailors' diets, attested to the need to provide significant amounts of high-quality, if monotonous, food. For extended voyages, the durability of food became a major concern. Aboard ship, the weekly ration per man averaged 7 pounds of biscuit, 7 gallons of beer, 6 pounds of meat, 2 pints of peas, 3 pints of oatmeal, 1/3 pounds of butter, and 2/3 pound of cheese (Dixon 1981, 176). The army's diet was less diverse. In 1816, the British Army Ration Scale per man per day was 1 pound of bread, 1 pound of fresh or salted meat, 1 pint of wine or 1/3 pint of spirits in lieu of wine (Crowdy 1980, 266). Unlike sea voyages, provisioning meat for armies could be accomplished either from local sources as needed or from more easily replenished supplies carried by the unit. If there were live animals in the vicinity of a campaign, fresh meat was provided; otherwise, the daily meat ration was salt meat. During long campaigns without fresh meat or vegetables, soldiers were expected to supplement their diet from the canteen by purchasing food from local sources. During the U.S. Civil War, "sutlers"—camp-following petty traders—were a common source of food supplies for soldiers (Lord 1969). Campaign-weary and mostly unpaid, soldiers, however, were hardly disposed to engage in market transactions with the local population. Poaching, stealing, or raiding thus were common avenues of military provisioning. As the size of armies grew, however, dependence on local sources could threaten the military objectives. It was not until 1922 that the British began to assume the full costs of provisioning its military (Crowdy 1980).

The provisions for the crews of British merchant ships were significantly worse than those of the navy. Disease aboard merchant ships was rampant. Some diseases, such as scurvy, had known cures, but the costs of prevention—fresh vegetables and fruits—were labor costs that ship owners saved at the expense of their workers' health (Drummond and Wilbraham 1939). Seamen's meals on merchant ships of Atlantic Canada were usually "hard bread, bootleg coffee, salt beef, and a greasy yellow broth called pea soup. Evenings we get a small taste of hash and soft bread but it's mighty small, I tell you" (Sager 1989, 227).

Early durable foods, such as salt meat, were an important factor in the growth of urban centers and in the ability of navies and merchant seamen to make extended voyages. Although the technique of salting clearly created a durable food out of such important protein sources as fresh meat and fish, the product's quality generally varied between the simply tolerable to the inedible. Salting, while a significant innovation, nonetheless was limited mostly to meat and fish. The demand for increases in the quality of diet came with increases in social stratification, especially a growing wage stratum in the urban centers (Hope 1990). It was the increasing size of the military, however, and the difficulties in importing foreign foods during war that were instrumental in the search for new industrial food technologies.

CANNING AND THE MAKING OF A MODERN INDUSTRIAL FOOD

Until the nineteenth century, the British military provisioned itself largely from the plunder of the lands, but the armies were becoming larger than the food available from plunder (Crowdy 1980). Napoleon's invasion of Russia, for instance, involved over a million soldiers, far more than could be provisioned from already impoverished local sources (Goody 1982). The French, then at war with most of the European governments and suffering destabilizing food scarcities from blockades, recognized the critical importance of developing new methods for provisioning their military as well as increasing the availability of nationally produced foods. The revolutionary French Directory offered a cash award of 12,000 francs for the discovery of a method of preserving foods. In 1809, Nicholas Appert, a French brewer and experimenter who had spent most of his life exploring food preservation methods, submitted a treatise[2] on hermetic sealing of food after working on the problem for a decade or more. His simple method was based on earlier practices (Goody 1982): Put food into a glass jar, stop the jar with a cork, and place the jar into boiling water for a time depending on what was in the jar. It was a major innovation with a tremendous practical success. France awarded Appert the prize after testing his product by shipping some of Appert's "preserved" foods "across the equator and far south, so as to subject them to extremes of temperature and the varying conditions of humidity and transportation" (Bitting 1937, 10). Appert's method not only revolutionized the technology for the preservation of meats but also created durable soups, milk, eggs, vegetables, fruits, essences, and many other products (Bitting 1937). Appert's method was *the* significant innovation leading to the creation of the first modern *industrial food*.

Appert's method originally used glass food containers that tended to break during their manufacture and transportation. It was the invention of the "tin" container that overcame this limitation. In 1810, Peter Durand was awarded a patent from the British government for the preservation of fruits, vegetables, and fish in sealed tinplate and glass cans, although the process was not widely adopted until 1820 (Bitting 1937). Glass jars were gradually abandoned because of their shipping weight and bulk and because they broke under the higher processing temperatures that were increasingly used.

It was the refinements to the tin technologies that lowered the costs of production and created a "wage food" (Friedmann 1990; Crouch and deJanvry 1980). The hand-made "plumb-joint" tin can was slow and costly to construct, 100 being a long day's work. In 1847, an Englishman, Allen Taylor, invented the stamp can, and two years later Henry Evan, Jr., of New Jersey invented the pendulum press for making can tops. In the late nineteenth century, Zimmerman invented the key-opening can which significantly reduced the cost of can manufacture and, at the same time, increased the convenience to the consumer. By 1900, can making itself had become a distinct industry and skilled trade and, although some canneries still made their own tins at the site of food canning, tin manufacture became increasingly an independent industry of the core sector. By 1890, the pace of production of the open-topped, pressed-out "sanitary can" had increased to the point that it was a fully machined process, with the United States the acknowledged leader in canning efficiency (Bitting 1937).

Appert's process and Durand's tins are the foundation of the development of modern industrial foods. Still, as with other inventions, many small refinements to the manufacturing process followed. For example, canners early learned that adding calcium chloride or potassium chloride to the water raised the processing temperature beyond the boiling point and decreased the processing time (Bitting 1937). Increasing the size and shape of the container in which the food was processed were further improvements. In 1820, the French Directory offered another prize for the preservation of "8 to 10 kilograms of animal substance in the same vessel for one year" (Bitting 1937, 13). The award was offered to further encourage wholesale manufacturing and to replace the small containers manufactured. Not only were small containers a hindrance on ships, but the round cans were difficult to store. Appert, who received the second prize in 1824, used tin cans with capacities from four to forty-five pounds and later developed the shapes of cans that are still in use today—round, oval, and rectangular. The tin cans could also be reused many times; when handles were added and the cans were reworked by sailors, the cans could be used as casseroles by the ship's cook (Bitting 1937).

Tinned foods, especially tinned meats, were rapidly adopted by the British Navy. Tinned meats were added to the navy's onboard medical supplies about 1815 as it was thought that, by replacing salted meats, the ravages of scurvy could be reduced (Drummond and Wilbraham 1939; Bitting 1937). In 1831, the Admiralty Regulations required tinned foods as part of the medical supplies on all its ships. In 1847, the Admiralty decided that tinned meats should be standard provision. Very soon after, however, the issue of quality arose again with tinned food, and the British Navy found itself condemning large quantities of tinned meat because of spoilage. In 1850, a Select Committee was appointed by Parliament to investigate. The committee concluded that the larger sized containers (9 to 14 pounds) were the ones most frequently going bad, a result of undercooking in the center of the can. The committee recommended that quality could be improved if the largest cans were limited to 6 pounds and if the Admiralty established its own cannery, which they did in 1856 (Drummond and Wilbraham 1939).

It was in the United States that canning first developed on a large commercial scale, although its innovators were English immigrants. Fishery products were the basis of nearly all the canneries that opened in the early nineteenth century. In 1819, hermetically sealed foods, principally seafood, were manufactured in New York City. In 1820, William Underwood and Charles Mitchell launched the canning business in Boston specializing mostly in pickles, sauces, jams, and mustard. In the 1830s, the canning business was established in Baltimore by canning oysters from the Chesapeake Bay area. Lobsters were canned in Maine in the 1840s and sardines followed in the 1870s; a turtle cannery was established in Florida in 1866. Salmon canning developed on the West Coast during the Civil War and rapidly became one of the most important seafood canning industries (Martin and Flick 1990; Cutting 1956; Goode 1887; Stevenson 1898).

Soon after the establishment of seafood canneries, fruits and vegetables also began to be canned, extending the number of durable foods available to the working class. These new items, which were at first canned in the seafood canneries during the slack season, soon established their own specialized canneries. The development of these canning industries in the United States played a role in how various industrial foods were incorporated into working-class diets. For example, with the use of tinned food during the Civil War, a relatively large market of postwar consumers was created.

CANNED FOODS IN URBAN INDUSTRIAL DIETS

Many of the core's military units had adopted canned foods by the mid-nineteenth century, and the merits of these foods in the provisioning of other large resident populations including prisons and educational institutions were increasingly recognized (Royal Dublin Society 1839). Canned meats began to appear in Britain in the late 1860s at prices half those of fresh meats and were consumed largely by the urban poor (Curtis-Bennett 1846; Johnston 1977; Tannahill 1973). By 1870, large quantities of tinned meat were being exported to Britain from the United States and Australia (Johnston 1977).

In addition to tinned meats, delicacies such as turtle soup, sauces, and pickles, continued to be packed in glass jars and were consumed by the wealthier classes (Ellis 1977; Royal Dublin Society 1839). But it was not until the late nineteenth century that canned foods other than these cheap meats became an integral part of working-class diets in Britain (Drummond and Wilbraham 1939; U.S. Tariff

Commission 1925). Until that time, most canned foods were expensive. Johnston notes that "contrary to a commonly held belief, tinned salmon at 9.5d for a small tin, was never an important item in the diet of the working class" (1977, 55–56). A description of tinned meat from 1874 gives some indication of the quality of the early tinned meats available and why these particular tinned foods were not as readily integrated into the diets of the poorest urban classes even if the food was ever "so cheap" (Horandner 1986; Drummond and Wilbraham 1939):

It was in a *big*, thick, clumsy red tin and was very cheap. . . . I have a vivid recollection of the unappetizing look of the contents—a large lump of course-grained lean meat inclined to separate into coarse fibres, a large lump of unpleasant looking fat on one side of it—and an irregular hollow partly filled with watery fluid. (Drummond and Wilbraham 1939, 382)

In the United States, you see a varied diet among urban and rural workers, in part due to the availability of industrial foods. Rural workers in the southern states continued to eat smoked and salted fish first introduced to the region during the eighteenth century when salted cod and dried and salted herring were exported from Newfoundland, the Gaspé Peninsula, and the Shetland Islands to the plantation sector in the southern United States and the Caribbean. The diets of rural sharecroppers in eastern Virginia consisted of meat from "low on the hog," usually, bacon, salt sides, trotters; lard; herring, both smoked and salted; corn meal; sugar; vegetables, including sweet potatoes, cabbage, and mustard greens; and, in some cases, dried fruit. There were also canned foods— blackberries, tomatoes, peaches, and pears (Frissell and Bevier 1899).

Studies of urban household diets in New York City show even greater variety. The diets of the working class included "animal foods"—bologna, bacon, corned brisket, smoked ham—along with some type of fish, including smoked herring, canned salmon, salted mackerel, canned sardines, dried and salted cod; milk; butter, and eggs; cereals, mostly bread, cakes, or rolls; sugar; vegetables, including onions, potatoes, canned and fresh tomatoes, canned corn, and cabbage; and a variety of fruits. Butter, milk, bread, and sugar were a part of virtually all household diets (Atwater and Bryant 1902).

Finally, a study of urban working women in Boston showed that women living in institutionalized settings, such as boarding houses and women's dormitories, had diets that were quite varied and that tinned foods were an important component in the meals prepared at the institutions. That study also found that increases in income coincided with an increase in the variety of foods in a woman's diet. Sweets, being one of the most expensive foods, was found most frequently among the higher wage earners, and preserved fish, salted, dried, or tinned, was a lunch food (Commonwealth of Massachusetts 1917).

Tinned foods had become a part of the U.S. merchant marine provisions by the end of the nineteenth century when the United States regulated the standard of provisions for maritime workers. In 1872, the provisions list included the following: bread (1 pound per day), beef (1-1/2 pounds on Sunday, Tuesday, Thursday, Saturday), pork (1-1/4 pounds on Monday, Wednesday, Friday), flour (1/2 pound on Sunday, Tuesday, Thursday), peas (1/3 pint on Monday, Wednesday, Friday), tea (1/8 ounce every day), coffee (1/2 ounce every day), sugar (2 ounces every day), and water (3 quarts every day) (Blume 1984, 42). This provisions list is not that much different from the British Naval rations of 1785 noted above. By 1898, industrial foods had been added, the provisions list included canned meat two days a week, fish on Fridays, potatoes/yams, canned tomatoes, peas, beans, rice, dried fruit, pickles, vinegar molasses, fresh bread and biscuits, corn meal, onions, lard, and butter (Blume 1984, 43).

The introduction of canned foods in Britain, however, did not alter the composition of the diets of the working class in the same way that it did in the United States. In Britain, working-class diets remained much the same as in earlier periods, although the quality of the foods appears to have improved. In 1901, a laborer's household in York had bread, butter or bacon, and sometimes porridge for breakfast; bread, meat or fish, and sometimes potatoes and cabbage for lunch; and bread, butter or ''dripping,'' and sometimes currant cake for tea; and if supper was served, bread and cheese or butter. The beverage was either coffee or tea (Rowntree 1902, 265–84). Higher waged workers, as in the early nineteenth century, had a more varied diet with more meat, vegetables, and sweets (pastry, jam pudding, rice pudding, cake, custards) with tea, coffee, and cocoa as beverages (Rowntree 1902, 185–89). As one can see, sugar was now firmly established in the diet of the core (Mintz 1985).

By World War I, tinned foods had become a significant part of the British diet with Britain the largest importer of canned foods (Johnston 1977). Another important industrial food during this period that was a precursor to the fast foods of the second food regime was the fish and chip. Fish and chips were another source of cheap food during the war (Walton 1992). With the increased world production of tinned foods during the war, the prices of these durable goods had fallen to the point that even the poor in Britain could afford them. After World War I, the British government organized a concerted effort to develop a British tinned food industry (Johnston 1977; Hudson 1978). By World War II, tinned foods were eagerly sought. Housewives would hoard their rationed flour coupons and exchange it for points that could be used to obtain tinned foods (Driver 1983).

CANNED SEAFOOD: DEVELOPMENT OF A GLOBAL COMMODITY

Preserved seafood has long been an important component of diets in the world-system. As one of the first durable foods, salted and dried fish provided an additional source of protein to early urban workers. Dried fish, because it was a higher quality durable food, was a luxury durable, a part of the diet of nobles (Tannahill 1973). Salted fish, in contrast, was easier to process than dried, but it also was the easiest type of durable protein food to spoil. If the fish used was not prime or if it was packed improperly it spoilt easily. The fish could be either salted too much or too little, and the packers' heels could damage the fish as it was being compacted in the barrel (Braudel 1979). Because it remained a "perishable" commodity, it was highly regulated in many European markets (Braudel 1979; Jenkins 1927). It appears that salted herring and salted cod did not necessarily compete. When the salted herring season was at an end, salted cod was just coming to market in northern Europe. The highly regulated market prevented out-of-season, lower quality herring from entering the market during cod season. Because herring was preferred to cod in some markets, even poor-quality herring sold at low prices could displace cod (Braudel 1979).

Tinning increased the durability of seafood, and canned seafood was a global food industry from its inception. Both salmon and sardines were first tinned in Europe (Stevenson 1898). The French built a sardine cannery in 1834, and "French canned sardines" soon became a luxury food in the United States. Although the Franco-Prussian War (1870–1871) interrupted the supply of French canned sardines to the United States, the French held their monopoly on the global market until 1880 when there was a "period of failure" from a scarcity of pilchard, the fish canned by the French (U.S. Tariff Commission 1925; Weber 1921).

Merchants in New York City began funding experiments in canning small herring during the Franco-Prussian War when naval blockades prevented the French canned sardines from reaching the New York metropolitan market. In 1875, the first cannery that tinned the small herring as "French sardines" was opened in Eastport, Maine. With the failure of the French pilchard fishery in the 1880s, Maine sardines soon became an important source of canned seafood. The two centers of global production in sardines until the early 1900s were Maine and France (Cowan and Thompson 1992; U.S. Tariff Commission 1925). Initially, Maine sardines were competitive with French sardines, but their quality declined during the late nineteenth century because of a shift in the canning ingredients, most notably the substitution of other oils for olive oil. By the early twentieth century, the quality had deteriorated to the point that Maine sardines were mostly consumed by the poor in the southern United States and in the eastern industrial cities.

Salmon was the other industrial seafood of the nineteenth century. The first canned salmon was produced in Scotland in 1824, but the scarcity of fish prevented the development of a canning industry there. In 1864, salmon canneries were established on the West Coast of the United States. The center of global production in salmon was the north Pacific with the United States and Canada the most important production centers. These two countries prepared 99 percent of the canned salmon in the 1890s, but this output had declined to 84 percent by 1928 as other countries entered the market (U.S. Tariff Commission 1929; Stevenson 1898). As with sardines, there was soon an international trade in canned salmon. South America and Australia were the first major importers with the British market growing strongly during the latter quarter of the nineteenth century (Stevenson 1898). Canned salmon also became an important component of the urban diet in the United States. Approximately 90 percent of U.S. production was consumed in country. In addition, the cheaper grades of Canadian salmon were sent to U.S. urban markets, while the more expensive grades of both the U.S. and Canadian packs were sent to Britain (U.S. Tariff Commission 1929).

By the early twentieth century, the United States, Canada, and Siberia produced canned salmon that was exported to Britain, Australia, New Zealand, the Philippine Islands, Belgium, France, Chile, Cuba, British Africa, Italy, and other countries (U.S. Tariff Commission 1929). Canned sardines were tinned in France, Spain, Portugal, Norway, the United States, Chile, India, Algeria, and Canada for export to British Africa, Egypt, Europe, Mexico, the West Indies, Central America, Brazil, Argentina, Chile, Peru, Venezuela, and other countries (U.S. Tariff Commission 1925). Sites of production and consumption remained relatively constant through World War II when frozen foods increasingly displaced major commodities that had heretofore been canned and consumed in the core countries (Scattergood 1952; Newell 1989).

CONCLUSIONS

Durability is a necessary condition for the establishment of a global food system. The barriers that prevent foods from entering into the global economy are broken in part by technological changes that enable foods to extend their life. In conclusion, we would like to draw attention again to the linkages between the various stages of food durability and their relation to a global food system and to what we think offers a useful way of conceptualizing food durability in historical context. We have discussed the linkages between different historical stages of durability from the late fifteenth century to the mid-twentieth century, emphasizing the importance of conceptualizing modern *industrial* foods as a bridging technology providing the transition from earlier forms of preservation to wholly manufactured foods characteristic of the second food regime. Durability itself is structured by changing technologies of preservation. As one moves from the first food regime to the second food regime, the range of durable food commodities increases, and there is an active reconstruction of durable foods with wholly new products created and consumed. Modern industrial foods, of which we regard canning the historical case, provided the bridging technology between simple preservation and these wholly manufactured foods of the second food regime.

The first linkage between durability in an emergent global food system is modern industrial preservation—the application of mass production technology to the creation of a durable food. The needs of the military, captured in Napoleon's recognition that "an army marches on its stomach," provided the impetus to develop more effective technologies of food preservation than the salted meats and fish so heavily relied upon by the armies and navies. The invention of the canning process by Nicholas Appert, offered as an award by Napoleon's own Directory, was the significant result. This was later to be followed by the introduction of tinned-plated containers to replace glass. Industrial processing not only provisioned the military, but also provided cheap meats and fish to an expanding urban population. While canned seafood was initially regarded as a luxury food of the wealthy, certain seafoods, such as sardines, rather quickly became cheap sources of meat protein. From their inception, these first forms of modern industrial foods were more durable than simple preserved foods.

A second important linkage between an expanding food system and a global economy was the introduction of metropolitan diets to the periphery. While the phenomenon of introducing the mass-produced food commodities of the urban metropole to the periphery is an important development in the second food regime, we think it important to recognize that those relationships between urban metropolitan diets and the periphery were also established at an earlier point through simple preservation and, later, canning. In the first food regime, canned foods enabled colonial administrators to maintain their metropolitan diet in the periphery. The local elite also became a ready market for these goods. Canned sardines and canned salmon, for instance, had a ready market in South America, British Malaya, the Philippines, British Africa, India, and Indonesia during the first food regime (U.S. Tariff Commission 1925).

A final linkage between durability and the global food system that requires elaboration is that between foods and foodstuffs and the industrial materials and technologies needed in their preservation/processing/manufacturing. For instance, tin mining, the making of tin plate, and the manufacture of cans were all necessary components for the production of canned foods. Another related linkage is suggested by the preservation technology possibilities of biotechnology and gamma irradiation and their relation to durable foods. For example, the potential for designing some degree of durability in fresh foods through the new biotechnologies is increasingly possible. This would suggest that the genetic resources themselves would be part of the commodity chain (Cowan 1987). There is, for example, substantial interest among food processors for altering the foodstuffs that they now tin or freeze. With biotechnology, food processors could enhance the quality of tinned and frozen food through designing some level of durability in the foods before they are preserved. General Foods, for instance, has expressed keen interest in designing broccoli that will not lose its deep green color or "mouth appeal" in the canning or freezing process. With new forms of durability stemming from alteration of the genetic constitution of fruits and vegetables, the durability continuum could actually circle back to fresh foods, leading to further integration of the global food system.[3] These emergent preservation technologies hold the potential for restructuring the durability continuum with new connections between wholly manufactured foods and fresh foods.

NOTES

The authors' names are simply listed in reverse alphabetical order. Equal contributions from each went into the paper and we share equal responsibility for the paper. We are grateful for the helpful comments of Philip McMichael, Clare Hinrichs, and Sidney Mintz.

1. A food regime is "the rule-governed structure of production and consumption of food on a world scale" (Friedmann 1993, 30; Friedmann and McMichael 1989). Two regimes have been elaborated: the first beginning in 1870 and ending in 1939, and a second one beginning in 1947 and ending with the food crisis of 1972–1973.

2. *Le Livre de Tous les Ménages ou l'Art de Conserver pendant Plusiers Années Toutes les Substances Animales et Végétales.*

3. For example, DNA Plant Technology Corporation, a U.S. agricultural biotechnology company, has recently announced the creation and successful marketing of a new proprietary tomato variety that stays fresh for from ten to fourteen days compared with most vine-ripened tomatoes whose shelf life is from three to seven days. The source of the extended shelf life is a patented breeding technique (*New York Times*, 6 August 1993, D3).

REFERENCES

Atwater, W. O., and A. P. Bryant. 1898. *Dietary Studies in Chicago in 1895 and 1896.* Office of Experiment Stations Bulletin 55. Washington, D.C.: U.S. Department of Agriculture.

———. 1902. *Dietary Studies in New York City in 1896 and 1897.* Office of Experiment Stations Bulletin 116. Washington, D.C.: U.S. Department of Agriculture.

Bitting, A. W. 1937. *Appertizing.* San Francisco: The Trade Pressroom.

Blume, Kenneth John. 1984. "The Hairy Ape Reconsidered: The American Merchant Seaman and the Transition from Sail to Steam in the Late Nineteenth Century." *The American Neptune* 44, 1: 33–47.

Braudel, Fernand. 1979. *The Structures of Everyday Life.* New York: Harper & Row.

Commonwealth of Massachusetts, State Department of Health. 1917. *The Food of Working Women in Boston.* Boston: State Printers.

Cowan, J. Tadlock. 1987. "An Emerging Structure of Technological Domination: Biotechnology, the Organization of Agricultural Research, and the Third World." *International Journal of Contemporary Sociology* 24, 1–2 (January–April): 31–44.

Cowan, J. Tadlock, and Susan J. Thompson. 1992. *The Spatial Organization of Extractive Industrialization: Seafood Processing in Maine, 1870–1980.* Paper presented at the annual meeting of the Rural Sociological Society, Pennsylvania State University, August 19.

Crouch, Luis, and Alain deJanvry. 1980. "The Class Basis of Agricultural Growth." *Food Policy* 5, 1: 3–13.

Crowdy, J. P. 1980. "The Science of the Soldier's Food." *The Army Quarterly and Defense Journal* 110, 3: 266–79.

Curtis-Bennett, Sir Noel. 1946. *The Food of the People.* London: Faber and Faber.

Cutting, Charles L. 1956. *Fish Saving.* New York: Philosophical Library.

Dixon, Conrad. 1981. "Pound and Pint: Diet in the Merchant Service 1950–1980." In

Charted and Uncharted Waters, edited by S. Palmer and G. Williams. Proceedings of a conference on the Study of British Maritime History, September 8–11, Queen Mary College, London. London: National Maritime Museum.

Driver, Christopher. 1983. *The British at Table, 1940–1980*. London: Hogarth Press.

Drummond, Jack C., and Anne Wilbraham. 1939. *The Englishman's Food*. London: Jonathan Cape.

Ellis, Audrey. 1977. *The Magpie History of Food*. London: Seven House Publishers.

Friedmann, Harriet. 1990. "Family Wheat Farms and Third World Diets: A Paradoxical Relationship between Unwaged and Waged Labor." In *Work without Wages*, edited by J. L. Collins and M. Gimenez, 193–213. Albany, N.Y.: State University of New York Press.

———. 1991. "Agro-Food Industry and Export Agriculture." In *Towards a New Political Economy of Agriculture*, edited by W. H. Friedland, L. Busch, F. H. Buttel, and A. P. Rudy, 65–93. Boulder, Colo.: Westview Press.

———. 1993. "The Political Economy of Food: A Global Crisis." *New Left Review* 197 (January/February): 29–57.

Friedmann, Harriet, and Philip McMichael. 1989. "Agriculture and the State System." *Sociologia Ruralis* 29, 2: 93–117.

Frissell, H. B., and Isabel Bevier. 1899. *Dietary Studies of Negroes in Eastern Virginia in 1897 and 1898*. Office of Experiment Stations Bulletin 71. Washington, D.C.: U.S. Department of Agriculture.

Garnsey, Peter. 1988. *Famine and Food Supply in the Graeco-Roman World*. Cambridge, England: Cambridge University Press.

Goode, George B. 1887. *The Fisheries and Fishing Industries of the United States*. Washington, D.C.: U.S. Commission of Fish and Fisheries.

Goody, Jack. 1982. *Cooking, Cuisine and Class*. Cambridge, England: Cambridge University Press.

Hall, Kenneth R. 1985. *Maritime Trade and State Development in Early Southeast Asia*. Honolulu: University of Hawaii Press.

Hope, Annette. 1990. *Londoners' Larder*. Edinburgh, Scotland: Mainstream Publishing Company.

Horandner, Edith. 1986. "Storing and Preserving Meat in Europe: Historical Survey." In *Food in Change*, edited by A. Fenton and E. Kisban, 53–59. Edinburgh, Scotland: John Donal Publishers.

Hudson, Kenneth, 1978. *Food, Clothes and Shelter*. London: John Baker.

Jenkins, James T. 1927. *The Herring and the Herring Fisheries*. London: P. S. King & Son.

Johnston, James P. 1977. *A Hundred Years of Eating*. Montreal, Quebec: McGill-Queens University Press.

Kenney, Martin, Linda M. Lobao, James Curry, and W. Richard Goe. 1991. "Agriculture in U.S. Fordism: The Integration of the Productive Consumer." In *Towards a New Political Economy of Agriculture*, edited by W. H. Friedland, L. Busch, F. H. Buttel, and A. P. Rudy. Boulder, Colo.: Westview Press.

Lord, Francis A. 1969. *Civil War Sutlers and Their Wares*. New York: Yoseloff.

Martin, Roy E., and George J. Flick. 1990. *The Seafood Industry*. New York: Van Nostrand Reinhold.

Mintz, Sidney W. 1985. *Sweetness and Power: The Place of Sugar in Modern History*. New York: Viking.

Newell, Diane. 1989. *The Development of the Pacific Salmon-Canning Industry*. Montreal, Quebec: McGill-Queen's University Press.

Ommer, Rosemary E. 1991. *From Outpost to Outport*. Montreal, Quebec: McGill-Queen's University Press.

Rediker, Marcus. 1987. *Between the Devil and the Deep Blue Sea*. Cambridge, England: Cambridge University Press.

Reid, Anthony. 1988. *Southeast Asia in the Age of Commerce 1450–1680*. New Haven, Conn.: Yale University Press.

Renner, H. D. 1944. *The Origin of Food Habits*. London: Faber and Faber.

Rowntree, Seebohm. 1902. *Poverty: A Study of Town Life*. 2d ed. London: Macmillan.

Royal Dublin Society. 1839. *Observations on the Preservation of Animal and Vegetable Substances*. Dublin, Ireland: Milliken and Son.

Sager, Eric W. 1989. *Seafaring Labour*. Montreal, Quebec: McGill-Queen's University Press.

Scattergood, Leslie W. 1952. *United States Imports and Exports of Herring and Sardines in Recent Years*. Research Bulletin 6. Augusta, Me.: Department of Sea and Shore Fisheries.

Smith, Richard. 1990. "The Historical Internationalisation of Food Processing in the Case of the Shetland Islands." In *Political, Social and Economic Perspectives on the International Food System*, edited by T. Marsden and J. Little, 215–29. Aldershot, England: Avebury.

Stevenson, Charles H. 1898. "The Preservation of Fishery Products for Food." *Bulletin of the U.S. Fish Commission* 18: 335–563.

Tannahill, Reay. 1973. *Food in History*. New York: Stein and Day.

U.S. Tariff Commission: 1925. *Sardines*. Washington, D.C.: U.S. Government Printing Office.

———. 1929. *Salmon*. Washington, D.C.: U.S. Government Printing Office.

Wallerstein, Immanuel. 1974. *The Modern World-System. Vol. 1, Capitalist Agriculture and the Origins of the European World-Economy in the Sixteenth Century*. New York: Academic Press.

———. 1980. *The Modern World-System II: Mercantilism and the Consolidation of the European World-Economy, 1600–1750*. New York: Academic Press.

Walton, John K. 1992. *Fish and Chips and the British Working Class, 1870–1940*. Leicester, England: Leicester University Press.

Weber, F. C. 1921. *The Maine Sardine Industry*. Bulletin 908. Washington, D.C.: U.S. Department of Agriculture.

PART II

Agriculture in World-Historical Perspective

4

Historical Transformations in Agrarian Systems Based on Wet-Rice Cultivation: Toward an Alternative Model of Social Change

Ravi Arvind Palat

The idea that there may be alternative technologies in itself implies the idea of technological pluralism in place of the until now almost universally accepted technological monism. In this case each social system and each political ideology, indeed each culture would be free to develop its own particular line. Why should there not be a specifically Indian technology alongside Indian art and why should the African temperament express itself only in music or sculpture and not in the equipment which Africans choose because it suits them better? . . . Might there not be an unmistakably Japanese technology, just as there are typically Japanese buildings and clothes?
—Robert Jungk[1]

Several scholars have recently challenged our understanding of the processes of long-term large-scale social change in fundamental ways by alleging that the search for the origins of capitalism in Europe is rooted in an ethnocentric bias deeply etched at the heart of the modern social sciences. Contending that the reigning orthodoxy is fatally flawed as it focuses solely on changes within Europe and is hence unable to assess whether these changes were themselves the effects of evolutionary processes elsewhere on the planet (Blaut 1992a, 357–58), the revisionists seek to widen their frame of inquiry to include the Eastern Hemisphere as a whole (excluding Australasia). Thus, observing the temporal contemporaneity of the evolution of exchange networks in many different parts of Africa, Asia, and Europe, they maintain that these should be perceived ''as nodes in a hemisphere-wide network or process of evolving capitalism'' (Blaut 1992a, 355; see also Abu-Lughod 1989, 1990; Amin 1988; Blaut 1992b). Hence, they argue that the roots of capitalism ought to be sought *not* in the transition from feudalism to capitalism in Western Europe, but "within a common international theatre of social and commercial changes" (Perlin 1983, 33). In short, capitalism from the start was hemispheric in scope and origin.

While all revisionists attribute the origins of capitalism to a hemisphere-wide nexus of commercial links and associated social changes rather than to developments within Europe, few would subscribe to David Washbrook's (1988, 76) assertion that "colonialism was the outcome of South Asia's own history of capitalist development." Janet Abu-Lughod, for instance, maintains that the rise of the West was predicated on a prior decline of the East, and of China in particular (1989, 1990). In a different vein, André Gunder Frank insists that there has been a *singular* world system since at least 2500 B.C. (and perhaps earlier) to the present day (Frank 1990, 1991a, 1991b, 1992; Gillis and Frank 1992), and that the task therefore is to trace the evolution of this system toward capitalism (cf. Wallerstein 1991, 1992). Finally, Jim Blaut argues that, while Africa, Asia, and Europe could be viewed as "landscapes of even development" prior to 1492, all progressing uniformly toward the goal of capitalist development, the European conquest of the Americas inaugurated a series of processes that enabled protocapitalists in Europe not only to dissolve feudalism in their region but also to overwhelm and subordinate rival protocapitalists elsewhere (Blaut 1992a; cf. Dodgshon 1992; Palan 1992).

Despite these differences between the various strands of revisionism, they are all agreed that, prior to 1492, there were no qualitative differences between Africa, Asia, and Europe. However, once it is granted that the seeds of capitalism were sprouting everywhere in the Eastern Hemisphere, capitalism is endowed with an aura of inevitability; it is cast as a product of the "natural" progression of the human race.[2] In this sense, these revisionist theories are virtually indistinguishable both from the modernization perspective which, in its Rostovian incarnation, universalized a misleading reading of British history (Rostow 1960) and the triumphant proclamations of capitalism as the culmination of human progress and the "end of history" by the ideologues of the right on the ruins of the socialist experiment in Eastern Europe and the Soviet Union.

By universalizing a model of sociohistorical transformation derived from the particular experience of northwestern European societies, the revisionists obliterate the distinctiveness of other historical social systems and deny the possibility of alternate patterns of social evolution. Rather than investigating the specific sociohistorical dynamics of the several distinct social systems in the Eastern Hemisphere, they assume that an expansion of exchange networks produces identical patterns of change everywhere. In effect, the revisionists attempt to transcend Eurocentrism by ruling out of court all the distinctiveness of non-European societies.

The present review of the patterns of historical evolution in societies based on wet-rice agriculture suggests that, despite the temporal contemporaneity of an expansion of relational networks in early modern Europe and Asia, the two processes were fundamentally dissimilar. Central to this argument are the fundamental differences in agricultural practices dictated by the dominant crops and the specific conditions of production in each area. Whereas the substitution of labor-power by animal and mechanical power represented technological progress in societies with low densities of population, the technical conditions of wet-rice cultivation dictated the substitution of simple tools for more complex instruments (Bray 1983, 4–5). This implies that, rather than moving toward large-scale consolidated farming operations, the dynamics of change in societies based on irrigated riziculture increasingly privileged small-scale operations. Or, as Thomas Smith (1980, 105) puts it so well, "To speak metaphorically, rather than impelling farming forward to a manufacturing stage of production, [operations associated with wet-rice agriculture] served to strengthen its handicraft character."

Once emphasis was placed on the skill of cultivators rather than on increasingly complex instruments of production, as was the case in early modern Europe, there was a tendential decline in the intervention of landlords in the production process. This implied that though producers may remain formally subordinate, no attempt was made by landowners to revolutionize and transform constantly the labor process. These conditions imposed severe impediments to a ceaseless accumulation of capital since landlords were unable to realize an increase in relative surplus value by constantly reducing production costs. At the same time, the increasing premium placed on skilled labor even constrained their abilities to realize an increase in absolute surplus value. There was hence no tendency toward an increasing real subsumption of labor to capital, identified by Karl Marx (1977, 1019–38) as the hallmark of capitalism as a mode of production sui generis.

In an attempt to delineate these tendencies, the first section of this chapter will outline some general features of wet-rice agriculture in medieval and early modern southern China and peninsular India and in Tokugawa Japan. This will be followed by an examination of the changes in agrarian relations, growth of regional and sectoral interdependencies, and transformations in the patterning of the relations of power in these areas. These sections will also briefly consider the very different historical experiences of the peoples inhabiting the region we now call Southeast Asia. Finally, the implications of this argument will be examined for the contemporary debate on long-term large-scale social change.

NATURAL CHARACTERISTICS OF IRRIGATED RIZICULTURE

Apart from a longer growing season, the ability of areas under irrigated rice agriculture to sustain substantially larger populations than the agrarian systems of early modern northwestern Europe derived from the natural characteristics of the rice plant. Since the staple crops of Europe (barley, rye, and wheat) "bore heads with relatively few grains—at best a few dozen, compared with several hundred grains in each panicle of rice or millet—and usually only a single head on each plant,"[3] a much higher proportion of the yield had to be reserved as seed for the next growing season.[4] Given these conditions, northern European agrarian systems were based on the extensive use of lands and were unable to sustain high population densities until comparatively recent times.[5] Finally, since an expansion of agrarian output was usually possible only by an extension of the arable, large farms had an advantage over small holdings because they could afford more draft animals and equipment and use these more efficiently (Kula 1976, 46; Bray 1983, 6–9; 1986, 199–202).

In contrast to the low yields of northern European staple crops, rice had a seed to yield ratio of approximately 1:100. Consequently, it was necessary to reserve a much smaller proportion of the crop for seed. Moreover, whereas the dominant staple crops of northwestern Europe drained nutrients from the soil, requiring constant manuring and necessitating leaving the fields fallow for at least one year in three until recently, rice obtained all the nutrients it required from water. This not only diminished the importance of livestock as a source of manure, but also meant that most fields could be kept in cultivation continuously.[6] In addition, Chinese and Indian historical sources indicate that for over two millennia, it was not uncommon for fields to yield two or three crops a year (Ho 1956; Perkins 1969, 43; Rawski 1972, 11ff.; Bray 1984, 491–95; Mukhia 1981, 288–89). Hence, lands under irrigated rice cultivation could support higher population densities as the area required for subsistence was much smaller than that required by the staple crops of northern Europe (Bray 1983, 9; Geertz, 1963, 32–33).

It was also possible to increase the productivity of lands under rice cultivation by greater control over the regulation of water supply, which often provided an alternative to an extension of the arable, especially since the latter course involved high investments of time and labor in preparing previously uncultivated land for agriculture (Geertz 1963, 36). However, since flooding was as much of a danger to the rice plant as drought, in the absence of mechanical equipment to level the fields, plots of land had to be small to ensure adequate control over drainage (Geertz 1963, 31; Gray 1983, 9, 12). Since it was easier to regulate water supplies more efficiently on small plots of land, significant increases in productivity could be achieved by decreasing the size of fields under wet-rice cultivation.

The progressive decrease in the size of plots with the construction of irrigation works led to a substitution of heavy iron hoes for animal-drawn implements. The cultivation of land with technologically simpler equipment was neither an indication of technological regression, nor of the material deprivation of the cultivators; it merely implied "that animal-drawn implements [were] not suited to such small-scale, highly-skilled farming techniques" (Bray 1984, 604; see also Geertz 1963, 95–101). In other words, the dimunitive dimensions of rice fields were dictated by the technical requirements of production rather than by the pressure of population on land.

At the same time, the great adaptability of rice, indicated by the wide range of its varieties, permitted farmers simultaneously to cultivate different varieties as an insurance against crop disease, pestilence, and the vagaries of the weather and to distribute their requirements of labor and water more optimally. The evolution of more effective techniques to regulate water supplies, along with the flexibility of seed selection, and an assortment of labor-intensive innovations (for example, planting seeds in rows to facilitate weeding and transplanting seeds sown in nursery beds)[7] led to the installation of multicropping regimes.

Thus, whereas technological progress in northwestern Europe was indicated by the growing use of labor-saving devices, agricultural improvements in rice-based agrarian systems were associated with labor-intensive techniques. The additional application of labor did not necessarily entail a corresponding decline in its productivity as the shift from single cropping to multiple cropping usually yielded an output far in excess of the increased inputs of labor necessary to accomplish the transition. Moreover, since labor requirements were spread throughout the year, household labor was usually sufficient to operate a small farm, and requirements for additional labor during peak periods were met by systems of cooperation which allowed communities of small cultivators to distribute available pools of labor more efficiently by staggering planting and harvesting operations (Bray 1986, 120–21; 1983, 12; Geertz 1963).

The distinctive characteristics of agrarian systems based on low-yield cereals and those based on high-yield cereals were historically expressed in the contrasting technical relations of production between the social systems of Europe and those in the rice-growing tracts of Asia. The potential of wet-rice farming to increase outputs with the additional application of labor meant that tendencies toward large-scale agricultural operations observed in Europe were virtually nonexistent in regions where rice was the staple crop as the costs of effective supervision of the numerous discrete tasks involved in its cultivation rose exponentially with no concomitant advantages. Indeed, the greater dependence of rice cultivation on the quality of labor, rather than on capital inputs, meant that a skilled small farmer was at least as capable as a wealthy landowner in raising the productivity of land. Or, as Francesca Bray wryly observes, ''[I]nspecting an irrigated field for weeds is almost an onerous as weeding it oneself'' (1983, 13; 1984, 604). Hence, while the economic calculations of large-scale agricultural enterprises led to a consolidation of demesne lands and an eviction of cultivators from their plots in late medieval and early modern Europe, no similar trend was inscribed in the historical record of rice-growing tracts of Asia. Land ownership in these societies, though less profitable than commercial cash-crop plantations, trade, or usury, was valued because it conferred social status and prestige and provided a greater security of investment.[8]

Since the productivity of fields could be achieved through additional inputs of labor, areas under wet-rice cultivation could support increasingly greater densities of population. The demographic growth made possible by intensive farming both facilitated an expansion in nonagricultural occupations and exerted a downward pressure on labor costs. It was only by the early eighteenth century, when population growth rates began to stabilize, that there was a secular rise in the cost of labor due to high wages in the sectors of commerce and craft production (Smith 1980, 105–6, 118; Gray 1986, 153).

Further, an important stimulus to the growth of machinery in Europe was the fact that wheat, barley, and rye flour could keep for several weeks without spoiling in temperate climates. Dehusked grains of rice and millet, however, could be kept for only a few hours. Thus, whereas feudal lords and wealthy burgers in medieval Europe constructed flour mills where peasant households could take several weeks' provisions of grain to be milled, rice had to be processed on an almost daily basis (Sigaut 1988, 5). Indeed, given the significantly larger densities of population in societies based on irrigated rice cultivation, there was no impetus to develop labor-saving machinery. It was this condition, rather than a "low-level equilibrium trap" (Elvin 1973), that stymied the development of metallurgy in China after the eleventh century.

Finally, it should be noted that although large-scale consolidated farming operations in wet-rice cultivation have been set up in the nineteenth and twentieth centuries in Australia and the United States, these enterprises were installed in areas with low population densities (Bray, 1983, 28), after the indigenous populations had been decimated and marginalized. Moreover, by the time of their installation, a specifically capitalist technology favoring the increasing use of labor-saving instruments of production had evolved. These developments do not, therefore, negate the main thrust of this chapter that tendencies promoting the ceaseless accumulation of capital were not present in areas dominated by wet-rice cultivation in late medieval and early modern Asia.

CHANGING PATTERNING OF RELATIONAL NETWORKS IN SOCIETIES BASED ON WET-RICE AGRICULTURE

These natural characteristics of wet-rice agriculture had several significant consequences for sociohistorical transformations in Asia. Most important, societies based on irrigated riziculture did not exhibit tendencies toward the creation of large, consolidated estates and the development of increasingly complex tools, since accumulation of capital did not bestow any special economic advantages. At the same time, the ability of fields under irrigated riziculture to support greater densities of population implied that a large, and constantly increasing, proportion of the population could be engaged in nonfood-producing activities on a full-time basis. The consequent elaboration of a sophisticated divisioning of labor and the attendant evolution of exchange networks were not confined to areas ecologically suited to wet-rice cultivation. Even in the more arid regions, cultivators and craftsmen began to specialize in the production of a variety of crops and goods, and they were linked to rice growers through long-distance trade dependencies. However, as well shall see, though the intensification of rice cultivation generated expansive circuits of exchange and promoted a range of artifactual production, these did not result in the creation of large-scale economic enterprises. Finally, the growing complexity of low-level transactional networks had profound implications for the patterning of power relations, particularly in large agrarian empires with high densities of population since the existing technologies of control proved incapable of catering to the routine administrative needs of the various parts of the empire. Consequently, unlike early modern European states which experienced a progressive centralization of power, large agrarian empires in areas based on wet-rice cultivation experienced a tendential devolution of centralized authority to regional units of administration or jurisdiction, even if many of these were larger in territory and population than most European states.

Changing Patterns of Agrarian Relations

Though the rising productivity of land in Song China (960–1280) made its purchase increasingly attractive to wealthy individuals, the tasks involved in intensive wet-rice cultivation simultaneously strengthened the hands of skilled cultivators in their dealings with landlords. Intensive cultivation actually increased the demands made on cultivators, unlike extensive cultivation with mechanical implements which led to a deskilling of the workforce. Prompt attention to weeding, careful selection of seeds, planting seedlings in rows, transplanting, and a multitude of other tasks all placed a premium on skill. In these circumstances, a larger labor force—with its mix of full-time and part-time workers, some of whom may have been indentured—was decidedly inferior to a tightly-knit, socially cohesive family farm which could rely on its members

to a far greater degree for spontaneous effort since it gave them stronger and more immediate incentives. Under the circumstances, technical innovations brought the opposite of the economies of scale we tend to mistakenly associate with *all* technological advance; that is, beyond a certain size, the larger the farming unit the more efficient it was likely to be. (Smith 1980, 105, emphasis in the original)

With the greater intensity of cultivation and the proportionate complexity and difficulty of effective supervision, tenurial relations shifted increasingly in favor of the tenants, leading eventually to permanent tenancy rights in many regions of irrigated rice farming (Rawski 1972, 17–20; Bray 1983, 19–20; 1984, 605–8).[9]

This combination of low returns on land ownership with high rewards for cultivation implied that really large holdings were uncommon in the rice-growing tracts of late medieval and early modern southern China. For instance, a study of patterns of land holdings in the seventeenth-century Yangtzi Delta indicates that really large estates of more than 10,000 *mou* (approximately 1,377 acres) were exceedingly rare; about 75 percent of the land was held by medium (100 to 500 *mou*, or 13.8 to 68.9 acres) and small (less than 100 *mou*) landowners (Huang 1974, 158; see also Chao 1986, 92–93; Golas 1980, 302–5).

Likewise agrarian historians have distinguished two distinct forms of land holding in late sixteenth- and early seventeenth-century Japan from the earliest extant cadastral surveys (*kenichich ö*) dating to 1583. In the Kinai region, where the Imperial Court and the large urban centers of Kyoto, Fushimi, Osaka, and Sakai were located, the early development of double cropping had resulted in the dissolution of large landed estates and the consequent constitution of a large number of small tenant holdings cultivated by family labor (Smith 1980, 2–5). In the rest of the country, however, where control over labor was crucial before the spread of intensive methods of cultivation in the eighteenth century, the predominant form of land holding was the large, centrally managed estate. These estates were worked partly by hereditary or indentured servants (collectively known as the *genin*) who were assimilated as junior members of the landowner's household and, more significantly, by a class of people who bore a remarkable resemblance to European serfs—generally known as the *nago*.[10]

With the unification of Japan under the Tokugawa shogunate in the early seventeenth century, and the consolidation of its power, the shoguns and their subordinate daimyos encouraged innovations in agricultural practices and an expansion of manufactures in a bid to increase their tax revenues. The opening up of communications networks after the establishment of the Tokugawa peace facilitated the dissemination of agricultural improvements from the more densely populated and intensively farmed regions to other areas. By the end of the century, treatises like the *Nōgyo Zensho* published in 1697, based on an intensive study of Chinese agriculture literature and careful experimentation in Japan, helped propagate more scientific methods of cultivation, seed selection, and other practical improvements (Smith 1980, 87–92; Bray 1983, 18; 1984; 609; Yamamura 1979, 284). Due to these measures, it has been estimated that the yield of lands under rice cultivation in the Kinai region increased by almost 75 percent between the early sixteenth and the early seventeenth centuries (Smith 1980, 99). At the same time, the increased availability of commercial fertilizers (oil cakes, dried fish, night soil collected from urban centers), owing to an expansion of circuits of communications and exchange, and the construction of irrigation works by the shogun, daimyos, and, from the early eighteenth century, by landlords, merchants, and prosperous cultivators as well, facilitated the spread of intensive wet-rice cultivation (Smith 1980, 92–93; Bray 1986, 150–52).

As a result of these measures, cultivators outside the Kinai region, who had hitherto subsisted primarily on such dry crops as barley and buckwheat and who grew rice only to pay taxes, now began to shift to the cultivation of rice, both for their own sustenance and for sale (Bray 1983, 18; 1984, 609). As in China, the greater complexity of tasks associated with wet-rice cultivation favored a trend toward smaller units of cultivation. This tendency was manifested both by the increasing attention devoted to tenurial relations in administrative handbooks after 1700 (Smith 1980, 5), and by a gradual transformation of the relations of production from the *nago* and other forms of hereditary and indentured servitude toward waged labor through a succession of intermediate forms (Smith 180, 108–39; Bray 1984, 610).

Parallel tendencies can be discerned from the more fragmentary evidentiary base of late medieval and early modern peninsular India. Noboru Karashima's pioneering study (1984, 3–35) of land transactions in the Chola empire (875–1279) in southern India indicates a progressive increase in transfers of land and a shift from the communal exercise of rights over land to individual dispositions of plots of land with the construction of irrigation projects. Moreover, while the growth of individual dispositions of plots of land were the most spectacular in areas of irrigated riziculture, such transactions actually registered a decline in areas of dry cultivation during the last century of Chola rule (Heitzman 1985, 145–46).

While this initial shift toward individual dispositions of plots of land appeared to suggest a concentration of large tracts of land in the hands of prominent individuals (Karashima 1984, 20, 27–31), the continued expansion of irrigation projects led to a greater elaboration of claims to the produce of land. In areas of irrigated riziculture, the rights of tenants were explicitly recognized in the inscriptional record, implying that land transfers did not adversely affect the cultivators of the soil (Palat 1988, 143–65). In the more arid and poorly endowed ecological zones, unsuited for rice cultivation, the patterns of agrarian relations were significantly different. As cultivation was possible only in these zones with the excavation of discrete tank-and-channel networks, the main agents for the extension of sedentary settlements were the tax farmers (*nayakas*) and imperial administrators who alone had the resources necessary to excavate these irrigation projects. Accordingly, in these submontane regions, large landowners (often tax farmers) appear to have leased cultivation rights themselves, implying a concentration of land and the consolidation of large estates producing a variety of cash crops (Palat 1988, 117–19, 160–61).

Though the growing intensity of wet-rice cultivation in medieval and early modern southern China and peninsular India and in Tokugawa Japan appear not to have strengthened the cultivators vis-à-vis the landlords, it must *not* be assumed that all areas suitable for irrigated riziculture experienced similar tendencies. This is particularly the case in much of modern-day Southeast Asia, where irrigated rice cultivation appears to have been confined to a few pockets in Java, Sulawesi, Pampanga, and to the flood plains of the major rivers (Reid 1980, 239; Watabe 1978; Takaya 1975; 1977; Ishii 1978; Lieberman 1991). On the flood plains of Burma and Thailand, technologies to control the seasonal floods were devised only by the early nineteenth century. Prior to that time, the unpredictability and intensity of floods precluded the development of transplanting techniques. Consequently, farmers were compelled to resort to broadcast sowing.[11] Moreover, since major portions of the fertile deltaic regions lacked natural levies, they were completely submerged during the monsoons and completely dried up during the dry season (Takaya 1977; 445).

In these areas where techniques of rice cultivation were less developed, relations of production were likely to resemble those prevalent in agrarian systems based on low-yielding crops since the inability of land to sustain large populations made the control of labor crucial for the continued reproduction of the patterning of social relations. In all these regions with low densities of population, opportunities for the development of a market in land were sharply curtailed, and cultivators were permitted access to their plots on more restricted terms than those prevalent in areas with high densities of population.

Growth of Sectoral and Regional Interdependencies

Increased yields of food grains made possible by intensive wet-rice cultivation also led to the growth of extensive sectoral and regional interdependencies and generated expansive networks of trade. By permitting increasingly larger proportions of the population to engage in nonfood-producing activities on a full-time basis, a rapid increase in agricultural output stimulated the growth of artisanal production. Indeed, it has been argued that the relationship between an intensification of rice cultivation and the proportion of households engaged in nonagricultural occupations in Ming China was so regular that it was "possible to use a logarithmic power curve formula to project the number of households in a district on the basis" of the number of cultivating households it contained (Hartwell 1982, 378–79).

As larger numbers of households became engaged in nonagricultural pursuits, the divisioning of labor became proportionately more complex with artisans specializing in ever narrower segments of the production process. For instance, while small rural cultivators in Japan had engaged in craft production in their homes with family labor, before the rapid expansion of wet-rice farming they had characteristically produced the requisite raw materials themselves and had sold the completed product at periodic markets. However, by the second half of the eighteenth century, craft production had been fragmented into a series of discrete operations, each performed by a separate household. Thus, the end user "stood at the end of a rather long series of market transactions" (Smith 1980, 79).

Similarly, the growing fragmentation of production processes in late medieval and early modern India is indicated by the increasing reliance of weavers on the purchase of thread either at periodic markets, or at the more permanent market centers, and in the progressively greater elaboration of a ritually ranked occupational segmentation of the population that we invoke by the idiom of caste (Palat 1988, 310–11). These tendencies were paralleled in south China where, from the eleventh century, weavers of silk textiles are reported to have depended increasingly on the market for their supplies of thread while cultivators of mulberry trees relied on exchange networks for their tools (Shiba 1970, 116–20).

The progressive fragmentation of production processes was marked by a corresponding ruralization of craft production and related changes in its organization (Palat 1991, 22–23). The employment of increasing numbers of households in artisanal production as a result of an intensification of wet-rice cultivation led to the emergence of a mass market in which price displaced quality and artistry as the primary consideration affecting production. In these circumstances rural producers had "a decisive advantage for they were less encumbered than urban producers by guild restrictions and were nearer to raw materials and water power. Moreover, their labor costs were far more elastic since they did not demand a livelihood from industry, merely part-time employment to fill the lulls in farming" (Smith 1980, 76; see also Bray 1984, 610). As a result, there was an irreversible locational shift of craft production from urban centers to rural areas, which was reflected in the depopulation of castle towns in eighteenth- and nineteenth-century Japan, despite considerable efforts by the Tokugawa shogunate to arrest this tendency (Smith 1988, 15–79). In late medieval peninsular India, too, a survey of tax schedules and other records indicate a shift of handicraft production, particularly to the submontane regions newly opened to sedentary settlements (Ramaswami 1985, 38–40; Palat 1988, 192–93, 208 note 10, 259 note 133; Palat and Wallerstein n.d.), from the environs of temple-centered urban places where it had largely been confined in the early medieval period (see Champakalakshmi 1986, 41–44; 1988).

While the concentration of craft production in small enclaves within urban centers in China before the eleventh century had facilitated official control over artisans, state-controlled workshops were progressively displaced as a source of supply to consumers by a widespread dispersal of silk production, organized on the basis of household labor. Thus, though state-controlled workshops for the production of textiles continued to operate in the twelfth century and beyond, they were increasingly relegated to supplementing taxes levied and collected in kind for provisioning the Imperial Court, including its requirements for conducting legal extra-jurisdictional trade (Shiba 1970, 111).

The locational shift of craft production from urban enclaves to the countryside vastly expanded the pool of artisans. In this context, it is instructive to recall Irfan Habib's (1969, 65) perceptive observation that, although Western visitors to early modern India routinely commented on how the inflexibility of the caste system prevented an optimal social allocation of labor, there is not a *single* reference to a shortage of labor in any branch of production. Under conditions of an abundant supply of labor, technological progress was denoted not only by the introduction of labor-saving mechanical devices but by the acquisition of greater degrees of manual dexterity by craftsmen specializing in ever narrower niches of the production process.

Apart from engendering a ruralization of craft production, an increased output of food grains also stimulated the production of a wide variety of goods (Shiba 1970, 112–14) and a growing regional specialization of production (Shiba 1970, 111; Bray 1984, 601–2). In the case of Tokugawa, Japan, it has been observed that increased yields of food grains fostered an expansion in the cultivation of nonfood crops, particularly cotton and sericulture, both centered in the Kinai region, though pockets existed in almost all provinces. One of the most striking instances of the growth of regional specialization comes from the poorly endowed Uonuma country in the Echigo province of northwestern Japan where rural cultivators had long occupied themselves by weaving the native grasses into coarse hemp fabrics during the harsh winter which precluded any possibility of double cropping. Weaving remained a subsidiary source of employment in this area until the latter part of the seventeenth century when techniques were perfected to bleach these textiles. The subsequent popularity of these coarse cloths increased so rapidly that local supplies of hemp were no longer sufficient to meet the demands of a swiftly expanding market, and producers had to depend on increasingly distant regions for the supplies of raw materials (Smith 1980, 77–80).

Likewise in China, a rise in agricultural productivity, as a result of improved yields and the greater ubiquity of multicropping of food grains, led to a growing regional specialization of production—not only in the manufacture of silk, but also in the cultivation of sugar in Szechwan, Kwangtung, and especially Fujian provinces, where it had eclipsed rice in several districts by the Ming era, and of tea, timber, oil seeds, and a variety of other crops elsewhere. This was reflected in the growth of a system of markets integrating localized circuits of exchange within progressively larger networks of circulation (Shiba 1970, 45–164; Elvin 1973, 164–78; Golas 1980, 298–99; Bray 1983, 17; 1984, 601–3; 1986, 127–28; Rawski 1972, 48 ff.; Skinner 1974, 1985). Similar tendencies are inscribed in the historical record of late medieval peninsular India where evidence indicates the increasing dependence of craftsmen and cultivators of a variety of cash crops in the arid interior on the rice growers of the fertile coasts and along the banks of the major rivers (Palat 1988, 283–318). Despite the rise of regional interdependencies, it is important to stress that the overwhelming majority of the population continued to depend primarily on local marketing networks for their needs, relying on purchases from distant regions for only a small proportion of their daily requirements.

However, these localized marketing networks must be seen as capillaries feeding the arteries of long-distance trade and eventually connecting small rural producers to distant customers. The chief agents in this integration of localized circuits within larger networks of circulation, drawing even poorly endowed ecological regions into the nexus of commercial relations, were the hundreds of thousands of petty peddlers who conveyed their merchandise from periodic market to periodic market on carts, pack animals, and on their heads (see Skinner 1974). Perhaps even more significant, these petty traders also dominated the networks of cross-oceanic trade. Since the fine silks and other items of luxury consumption, from which a substantial share of profits were derived, catered only to small clienteles and occupied very little space, even wealthy traders who could outfit a vessel on their own were compelled to lease space aboard their ships to small traders who had acquired detailed knowledge of customer preference through networks of caste and kinship (Palat 1988, 288–89, 307–10).

By continually reinforcing the highly style- and pattern-specific localized markets, the presence of small traders in extralocal circuits of exchange proved inimical to the establishment of large, centralized mercantile enterprises (Palat 1991, 27). A more important impediment to the hierarchical subordination of craftsmen by wealthy merchants stemmed from the very organization of craft production. The ruralization of manufacturing operations, in the context of a relative abundance of labor, meant that it was easier to increase production by expanding employment rather than by installing new forms of labor control or by innovating production processes (Palat and Wallerstein n.d.).

The patterning of sectoral and regional interdependencies engendered by intensive wet-rice cultivation was predicated on a high degree of monetization of economic relations. Though low-level transactions were conducted with copper, tin, and lead coinages—as well as a variety of baser currency media including cowry shells and *badams* (an inedible, bitter almond)—revenue claims had to be met with gold and silver coins. Consequently, tax obligations were discharged by exchanging copper and other low-value currencies for gold and silver coins through the market mechanism. In these circumstances, unless there was a steady influx of precious metals, particularly to the bullion-starved economies of India and China, there would be an increase in the real rate of taxation (by raising the copper price of silver, for instance) and a contraction of economic activity. It was this importance of the inflows of bullion to the continued patterning of relational networks that stimulated an expansion of circuits of exchange originating in China to Southeast Asia and to Japan (Atwell 1986, 1990). Under similar imperatives, a contemporaneous expansion of patterns of circulation in peninsular India tied the coasts of the subcontinent to West Asia and Southeast Asia.

Neither the evolution of these long-distance networks of trade, conveying impressive volumes of low-value commodities, nor a recognition of the importance of inflows of bullion to the constitution and reconstitution of relational networks in the large agrarian empires of India and China implies that they were constituent elements of a singular historical social system. The "southern ocean" remained a secondary source of bullion for China, which obtained most of its bullion from Japan at least until the mid-eighteenth century (Atwell 1986, 1990), as suggested by the frequent bans of foreign trade to entrepôts on the Malay Peninsula and the eastern Indian Ocean archipelago (see Elvin, 1973, 216–20). However, trade across the Bay of Bengal was vital to the continued reproduction of societal networks for the peoples living along the coasts of peninsular India and their related hinterlands (Palat 1988, 1991; Palat and Wallerstein n.d.). Thus, there were at least two distinct and autonomous divisionings of labor—one centered in China and the other in the South Asian subcontinent—meeting at the busy bazaars of Melaka and Aceh.

Changes in the Patterning of the Networks of Power

The installation of multicropping regimes and the associated demographic growth and increasing density of low-level transactional networks also had far-reaching consequences for the patterning of the networks of power relationships in the areas where wet-rice cultivation was dominant. Since the seed to yield ratios of lands were at least twenty-five (and possibly even a hundred) times greater than those under dry-gain crops, it theoretically permitted correspondingly higher magnitudes of surplus extraction. Moreover, subsistence requirements appear to have been significantly lower in tropical regions than in subtropical and temperate zones (Habib 1978–1979). Even if a substantial portion of this surplus was devoted to the maintenance of a large segment of the population as retainers of the elites, the superior productivity of agriculture and lower subsistence requirements implied that political authorities in rice-growing areas were not dependent on the accumulators of capital for their protection-providing activities, unlike the case in Europe (for which, see Tilly 1990, 58–61). Thus, for instance, instead of relying on loans or cash advances from urban patriciates to wage wars or to suppress local rebellions, commanders of imperial forces in late medieval and early modern China and India merely drew cash from provincial treasuries to pay the troops under their command.[12]

While the Tokugawa shoguns were similarly freed from dependence on urban patriciates for their protection-providing services, especially since they did not have to finance protracted campaigns against well-entrenched rebel forces since the institution of the *sankin-kōtai* (alternate attendance) system significantly reduced the propensity of the daimyos to rebel. At the same time, the *sankin-kōtai* system meant that the daimyos were often dependent on loans from wealthy merchants since they were compelled to maintain residences and large retinues both at the shogunal capital and at their country seats.

Although the greater productivity of agriculture made political operatives in late medieval and early modern China and India less reliant on urban patriciates than European potentates or the daimyos, the intensification of regional and sectoral interdependencies and rising populations associated with wet-rice farming also steadily undermined the ability to centralized political apparatuses to control effectively the routine administrative needs of the several parts of the various empires. The growing complexity of low-level transactional networks and the enormous territorial extent of densely populated empires gradually outstripped the capacity of the communications system to sustain centralized control over their claimed jurisdictions. Simultaneously, units of local administration were equally unable to police transactional networks that increasingly spanned their boundaries. Consequently, larger political entities either retained a nominal existence with effective control passing into the hands of powerful regional satrapies, as in the case of post–T'ang China, or else split up into several autonomous polities, each defined by their participation in a system of power relations, as in the case of late medieval and early modern India.

The administrative history of the agrarian empires of China after the rebellion of An Lu-shan (755–763) clearly reveals this shift in the balance of power from the centralized political apparatus toward large regional satrapies. On the one hand, the growing densities of population strengthened the hands of district (*hsien*) magistrates who were conferred with the powers to collect taxes, settle disputes, and maintain law and order, and the concomitant eclipse of the pre-fectures (the *chou* or *fu*), or intermediate levels of government which had ex-ercised these functions previously. On the other hand, the increasing intensity of low-level transactional networks demanded a degree of coordination that nei-ther the district magistrate nor the distant imperial bureaucrats could provide. Consequently, after a series of ad hoc measures, a new level of regional admin-istration, the *sheng* or province, had evolved by the time of the early Ming. The importance of these large regional administrations was reflected in a major change in the career patterns of officials: rather than making interregional trans-fers within specific administrative branches, they were now moved between diverse fields within the same region. Finally, the growing ascendancy of re-gional units of administration consolidated the position of the local gentry and led to the virtual elimination of a professional bureaucratic elite.[13] Henceforth, diverse regional factions hurled accusations of ideological heterodoxy at their opponents in empire wide conflicts to protect their specific, localized interests (Hartwell 1982, 394–425).

An expansion of the circuits of exchange also undermined the existing pat-terning of political relationships in the early medieval South Asian subcontinent. While the growing complexities of low-level patterns of circulation subverted the territorial coherence of administrative units on which early medieval polities had been based, the spatial expansion of relational networks led to the formation of alliances across existing jurisdictional boundaries by opposing coalitions of nascent classes—as indicated by the rise of corporate bodies of itinerant trading groups and dominant landowners, on the one hand, and of subordinate artisanal and cultivating castes, on the other (Palat 1988, 84 ff.). The profession anach-ronism of early medieval political linkages was eventually overcome only by the transmission of new administrative procedures and practices from West Asia. The central elements of the new political dispensation was the introduction of prebendal forms of revenue assignments (the *iqtä'* and *nayankara* systems) whereby rulers attempted to secure reliable supplies of armed troops and money by temporarily assigning revenue claims from specified areas to their chief sub-ordinates. The widespread adoption of this practice, by several aspirant state builders, led to the evolution of a system with multiple autonomous polities marked by the periodic ascent of one of them to short-lived dominance (Palat and Wallerstein n.d.). Despite a façade of centralized control within these pol-ities, the practice of revenue assignments merely entrenched the power of re-gional potentates within their localities.[14]

In sharp contrast to the decline of centralized bureaucratic apparatuses in China and India, the Tokugawa shogunate experienced no similar tendency largely due to its more compact territorial extent, smaller population, and an absence of serious external threats (Atwell 1986, 230). Finally, the small prin-cipalities of the eastern Indian Ocean archipelago and the Malay Peninsula stood at the opposite end of the spectrum from the large agrarian empires of China and India. Since the political fortunes of Southeast Asian city-states were integrally linked to trade, there was a steady increase in the importance of the mercantile aristocracy, the *orang kaya*, which often acted as a powerful check on royal absolutism (Kathirithamby-Wells 1987, 31). Even when the sultans of Aceh in Sumatra began to intervene actively in the networks of circulation in the sixteenth century to counter Portuguese naval harassment, they continued to work in concert with the mercantile aristocracy. However, after the sultans negotiated a treaty with the Portuguese at the turn of the century, they used their increased coercive power to subdue the powerful *orang kaya* (Reid 1980, 240). Thus, the decline of centralized political authority in India and China was matched by a rise in absolutism in Japan and Sumatra.

THEORETICAL IMPLICATIONS FOR THE STUDY OF LONG-TERM LARGE-SCALE SOCIAL CHANGE

> In every general conjuncture, different countries react differently, whence the inequalities of development which, in the end, make history.
>
> —Pierre Vilar[15]

This survey of the processes of sociohistorical evolution in late medieval and early modern southern China and peninsular India and in Tokugawa Japan has indicated that the evolution of exchange networks in societies based on wet-rice cultivation led to a pattern of large-scale social change that was fundamentally different from the processes of transformation in early modern northwestern Europe. The high premium placed on skilled labor in the former privileged small-scale family farms rather than large-scale, consolidated agrarian enterprises. Similarly, the ability of fields under irrigated riziculture to support high densities of population reinforced the handicraft character of artisanal production and did not lead to a deskilling of the labor process. Thus, although the intensification of wet-rice agriculture stimulated a growth of sectoral and regional dependencies, and the integration of localized circuits of exchange within larger patterns of circulation, the accumulation of capital did not endow entrepreneurs with competitive advantages. Hence, the chronological simultaneity of an expansion of exchange networks in early modern Europe and Asia cannot be taken as indicative of a hemisphere-wide evolution of capitalism as the revisionists would have us believe.

The revisionist celebration of capitalism as the teleological goal of *all* human history is, in fact, symptomatic of a virtual paralysis in socialist thought after the collapse of communism. The attendant adulation of the market ignores the fact that markets are rooted in particular social and political contexts—that the presence of markets does not necessarily create price-making institutions (see Polanyi, Arensberg, and Pearson 1957). Their failure to excavate the causes that engendered the growth and expansion of market networks in Asia leads them to unwarrantedly assume that the stimulus toward an intensification of commercial relations was the same all across the Eastern Hemisphere. The consequent neglect of the structures of production renders the pattern of technological evolution in the wet-rice growing societies of Asia inexplicable in revisionist accounts. Finally, by reducing the processes of capitalism to the evolution of exchange networks, they obscure the specificity of the European experience itself and cannot account for the predatory nature of the capitalist world-system.

In contrast, this survey of the patterns of historical transformation in agrarian systems based on wet-rice cultivation suggests that impediments placed on the accumulation of capital by the natural characteristics of irrigated riziculture led to the extensive development of petty commodity production without an accompanying real subsumption of labor characteristic of the northwestern European experience. Without the development of large consolidated agriculture enterprises or the subordination of artisans to large mercantile corporations, the integration of production processes in Asia could not lead to colonialism and imperialism.

It should be noted that a recognition of the determinative impact of intensive wet-rice agriculture on the processes of sociohistorical evolution in late medieval and early modern southern China and peninsular India and in Tokugawa Japan does not entail the proposition that all these societies developed along identical lines. Indeed, there were significant differences in the patterns of change, representing variations of a similar process.

While the utility of comparative history lies precisely in its ability to compare and contrast developments in more than one unit of analysis, a finely textured analysis must also seek to recover the concrete specificity of each historical system. In short, Eurocentric conceptions of long-term large-scale social change can be purged from world historical studies only if we can recover the particular dynamics of change in historical social systems outside the privileged arena of Europe before they were subordinated to the drives of the capitalist world-economy.

NOTES

I acknowledge the helpful comments of Walter Goldfrank and Philip McMichael on earlier drafts of this chapter.

1. Cited in Alvares (1991, 208).

2. "Capitalism would (one suspects) have arrived in any case, but it would have arrived many centuries later and it would not have seated itself in Europe alone (or first) had it not been for European colonialism in America" (Blaut 1992a, 356). Or, as Blaut says elsewhere, "If the Western Hemisphere had been more accessible, say to South Indian centres than to European centres, then very likely India would have become the home of capitalism, the site of a bourgeois revolution, and the ruler of the world" (1992a, 369).

3. See Bray (1983, 5; 1986, 15, 198). According to Slicher van Bath's calculations, medieval northern European seed to yield ratios for barley, rye, and wheat were between 1:3 and 1:4 (Bray 1983, 28, note 2; 1986, 225, note 2). On the fields of the bishopric

of Winchester, Georges Duby (1981, 343) estimates that seed to yield ratios rose from 1:4.22 in 1300–1349 to 1:4.45 in 1400–1449 for wheat, from 1:3.8 to 1:4.31 for barley, and from 1:2.42 to 1:3.62 for oats. Guy Bois (1984, 205) calculates that this ratio was 1:6 on the very best lands in late fifteenth-century eastern Normandy, and slightly lower on less fertile soils. These figures are compatible with Emmanuel Le Roy Ladurie's estimates in which, however, he criticizes Slicher van Bath's finding that the average French yields rose to between 1:6.8 and 1:6.9 for the years 1500 to 1700 as highly excessive. Le Roy Ladurie argues, instead, that yield ratios in the range of 1:4 and 1:5 are more plausible (Le Roy Ladurie 1987, 113–15).

Though even plants like wheat, barley, and rye could theoretically produce a maximum of approximately 400 grains per plant as early as sixteenth-century Europe, the physiology of the plant and inefficient sowing techniques reduced the seed to yield ratio to about 1:5. The transplantation techniques developed in Asian rice cultivation (on which see below) however meant that rice had a much higher ratio to about 1:100 (Bray 1986, 5).

4. Even at the turn of the eighteenth century, a French peasant had to reserve from 15 to 20 percent of the crop as seed for the next season (Goubert 1986, 203).

5. For instance, according to the Domesday Book, an average allotment of land to a serf in eleventh-century England was approximately thirty acres. In contrast, an average Chinese cultivating household of that time worked less than five acres of land (Chao 1986, 222).

6. "The distinctive feature of wet-rice fields is that, whatever their original fertility, several years' continuous cultivation brings it up to a higher level which is then maintained almost indefinitely. This is because water seepage alters the chemical composition and structure of the different soil layers in a process known as *pozdolization* [*sic*]" (Bray 1986, 28, italics in the original; see also Geertz 1963, 29–31).

7. Though irrigation projects facilitated multiple cropping, the natural growth period of the rice plant ranged from six to seven months. One important way to overcome this limitation—known to Chinese, Vietnamese, and Indian cultivators for some two thousand yeas—was to reduce the time plants spent in irrigated fields by sowing seeds in well-manured nursery beds, and then transplanting the seedlings once they were about from 20 to 25 centimeters high.

The earliest reference to transplanting rice in Chinese literature is found in *Ssu Min Yüeh Ling*, a text of the second century A.D. (Bray 1984, 519). For transplanting in China and Vietnam, see Bray (1984, 501–4); for India, see Mukhia (1981, 288) and Alayev (1982, 227).

Transplanting also ensured that seedlings could be planted at regular intervals, thereby facilitating weeding. It has been estimated that transplanting increased yields by approximately 40 percent over broadcast sowing (Bray 1983, 11; Geertz 1963, 35, 77–78).

8. Or, as Bray succinctly puts it, "[o]ne did not make a fortune through being a landlord, one became a landlord through making a fortune" (1983, 30; 1984, 608). See also Golas (1980, 302).

9. A specialized case of permanent tenancy rights was the *yi tian liang zhu* (two owners of a single field) system under which, on payment of *fen-tu yin* (manured field silver), cultivators received transferable and negotiable rights over the topsoil that they could sublease or sell without the consent of the landowner. The sale of either one of these rights—the cultivation rights or the rights to the topsoil vested in the tenant, and the landowner's rights to the subsoil—did not affect the continuance of the others in any way. This system further strengthened the position of tenants since the owners of the

subsoil typically resided in distant urban centers and often did not know the precise location of their plots. A system that permitted a division of the rights to the topsoil and to the subsoil, and allowed these to be transferred separately, created considerable confusion in tax records. These complications were multiplied when the landowner, who bore the responsibility for taxation, sometimes farmed out the burden of collecting rents and paying taxes to a third party—the "three lords to a field system" (*i thien san chu*) of the Chang-chou prefecture being a case in point. Indeed, most of the information about these changing tenurial conditions are available to modern-day historians from the administrative records of a bureaucracy determined to curb tax evasions (Rawski 1972, 19–24).

10. Despite local variations in nomenclature and status, the *nago* generally lived separately from the landowners and held small plots of land. However, they held these plots, often barely adequate for their subsistence, as allotments from the landowners to whom they owed labor services. Moreover, though the *nago* tilled their plots and may even have paid the taxes assessed on their plots, legal responsibility for the payment of taxes rested on the landowners. Their lack of revenue responsibilities denied the *nago* access to village commons and waste lands and rendered them ineligible to participate in the deliberations of village assemblies (Smith 1980, 8–10). It has often been suggested that the plots allotted to the *nago* were often so infertile and small that they were dependent on the landowners even for their subsistence requirements (Yamamura 1979, 285).

11. "Rice is carefully transplanted in the inter-mountain basins, but it is broadcast haphazardly in the deltas. Broadcasting is adopted because there is not enough time for transplanting. No one can predict exactly when the flood will arrive; when farmers realize the flood is comming [*sic*] near, it is already too late for them to prepare for transplanting. The flood submerges all the delta in a very short time. Under such conditions, all the farmers can do is to broadcast seends [*sic*] on uninundated fields well in advance of the flood so that the seeds, using moisture from ephemeral showers, germinate and grow tall enough to survive the flood when it arrives" (Takaya 1977, 445). For Burma, see Lieberman (1991, 8–11).

12. The extraction of large agrarian surpluses in the Mughal empire led to a similar patterning of power relationships (see Richards 1990, 628). Though wet-rice cultivation was not dominant in the heartland of that northern Indian polity, the empire was an integral element of an emerging historical social system centered around, and integrated by transport across, the Indian Ocean coasts and their related hinterlands. Constraints of space preclude a discussion of the patterns of sociohistorical evolution in regions adjacent to zones where wet-rice agriculture predominated.

13. It has often been held that civil service examinations offered an avenue for upward social mobility in imperial China (see Ho 1962). However, after an exhaustive analysis of the career patterns of more than five thousand top officials, Robert Hartwell (1982, 419) concludes that there "is not a single documented example . . . of a family demonstrating upward mobility solely because of success in the civil service examinations. Indeed, in every instance of upward mobility supported by literary evidence, passage of the test *followed* intermarriage with one of the already established elite gentry lineages" (emphasis in the original).

14. Even in the heartland of the most centralized of Indian polities—the Mughal empire—where regional officials were frequently transferred, they continued to develop local bases of support as their reassignments were often within the same region. More-

over, in many instances, officials were reappointed to their former posts after short intervals (see Singh 1988).

15. Cited in Wallerstein (1980, 224).

REFERENCES

Abu-Lughod, Janet. 1989. *Before European Hegemony: The World-System* A.D. *1250–1350*. New York: Oxford University Press.

———. 1990. "Discontinuities and Persistence: One World-System or a Succession of Systems?" Paper presented at the conference on Early Modern Empires and Economies, Harvard University, Cambridge, Massachusetts, March 28–30.

Alayev, L. B. 1982. "The Systems of Agricultural Production: South India." In *The Cambridge Economic History of India*, vol. 1, *c.1200–c.1750*, edited by T. Raychaudhuri and I. Habib, 226–34. Cambridge, England: Cambridge University Press.

Alvares, Claude. 1991. *Decolonizing History: Technology and Culture in India, China and the West, 1492 to the Present Day*. Goa, India: The Other India Press.

Amin, Samir. 1988. *Eurocentrism*. New York: Monthly Review Press.

Atwell, William S. 1986. "Some Observations on the 'Seventeenth-Century Crisis' in China and Japan." *Journal of Asian Studies* 45, 2: 223–44.

———. 1990. "A Seventeenth-Century 'General Crisis' in East Asia?" *Modern Asian Studies* 24, 4: 661–82.

Blaut, J. M. 1992a. "Fourteen Ninety-Two." *Political Geography* 11, 4: 355–85.

———. 1992b. "Response to Comments by Frank, Amin, Dodgshon, and Palan." *Political Geography* 11, 4: 407–12.

Bois, Guy. 1984. *The Crisis of Feudalism: Economy and Society in Eastern Normandy*. Cambridge, England: Cambridge University Press.

Bray, Francesca. 1983. "Patterns of Evolution in Rice-Growing Societies." *Journal of Peasant Studies* 11, 1: 3–83.

———. 1984. *Agriculture*. In *Science and Civilisation in China*. vol. 6, *Biology and Biological Technology*, edited by J. Needham. Cambridge, England: Cambridge University Press.

———. 1986. *The Rice Economies: Technology and Development in Asian Societies*. Oxford, England: Basil Blackwell.

Champakalakshmi, R. 1986. "Urbanisation in Medieval Tamil Nadu." In *Situating Indian History (for Sarvepalli Gopal)*, edited by S. Bhattacharya and R. Thapar, 34–105. Delhi: Oxford University Press.

———. 1988. "The Urban Configurations of Tondai Mandalam: The Kanchipuram and Madras Regions, c. A.D. 600–1300." School of Social Sciences Working Paper Series. New Delhi: Jawaharlal Nehru University.

Chao, K. 1986. *Man and Land in Chinese History: An Economic Analysis*. Stanford, Calif.: Stanford University Press.

Dodgshon, Robert A. 1992. "The Role of Europe in the Early Modern World-System: Parasitic or Generative." *Political Geography* 11, 4: 396–400.

Duby, Georges. 1981. *Rural Economy and Country Life in the Medieval West*, trans. by Cynthia Postan. Columbia: University of South Carolina Press.

Elvin, Mark. 1973. *The Pattern of the Chinese Past*. Stanford, Calif.: Stanford University Press.

Wet-Rice Cultivation • 75

Frank, André Gunder. 1990. "A Theoretical Introduction to 5,000 Years of World-System History." *Review* 13, 2: 155–248.

———. 1991a. "Transitional Ideological Modes: Feudalism, Capitalism, Socialism." *Critique of Anthropology* 11, 2: 171–88.

———. 1991b. "A Plea for World System History." *Journal of World History* 2, 1: 1–28.

———. 1992. "Fourteen Ninety-Two Once Again." *Political Geography* 11, 4: 386–93.

Geertz, Clifford. 1963. *Agricultural Involution: The Processes of Ecological Change in Indonesia*. Berkeley: University of California Press.

Gillis, Barry K., and André Gunder Frank. 1992. "World System Cycles, Crises, and Hegemonial Shifts, 1700 BC to 1700 AD." *Review* 15, 4: 621–87.

Golas, P. J. 1980. "Rural China in the Song." *Journal of Asian Studies* 39, 2: 291–325.

Goubert, Pierre. 1986. *The French Peasantry in the Seventeenth Century*, trans. by I. Patterson. Cambridge, England: Cambridge University Press.

Habib, Irfan. 1969. "Potentialities of Capitalistic Development in the Economy of Mughal India." *Journal of Economic History* 29, 1: 32–79.

———. 1978–79. "Technology and Barriers to Social Change in Mughal India." *Indian Historical Review* 5, 1–2: 152–74.

Hartwell, Robert M. (1982). "Demographic, Political, and Social Transformations of China, 750–1550." *Harvard Journal of Asiatic Studies*, 42, 2: 365–422.

Heitzman, E. James. 1985. "Gifts of Power: Temples, Politics, and the Economy in Medieval South India." Ph.D. diss., University of Pennsylvania, Philadelphia.

Ho, Ping-ti. 1956. "Early-Ripening Rice in Chinese History." *Economic History Review* 2d series, 9, 2: 200–218.

———. (1962). *The Ladder of Success in Imperial China*. New York: Columbia University Press.

Huang, Ray. 1974. *Taxation and Governmental Finance in Sixteenth-Century Ming China*. Cambridge, England: Cambridge University Press.

Ishii, Yoneo. 1978. "History and Rice-Growing." In *Thailand: A Rice-Growing Society*, edited by Y. Ishii, trans. by P. Hawkes and S. Hawkes, 15–39. Honolulu: University of Hawaii Press.

Karashima, Noboru. 1984. *South Indian History and Society: Studies from Inscriptions, A.D. 850–1800*. Delhi: Oxford University Press.

Kathirithamby-Wells, J. 1987. "Forces of Regional and State Integration in the Western Archipelago, c. 1500–1700." *Journal of Southeast Asian Studies* 18, 1: 24–44.

Kula, Witold. 1976. *An Economic Theory of the Feudal System: Towards a Model of the Polish Economy, 1500–1800*, trans. by L. Garner. London: New Left Books.

Le Roy Ladurie, Emmanuel. 1987. *The French Peasantry, 1450–1650*, trans. by A. Sheridan. Aldershot, England: Scolar Press.

Lieberman, Victor. 1991. "Secular Trends in Burmese Economic History, c. 1350–1830, and Their Implications for State Formation." *Modern Asian Studies* 25, 1: 1–31.

Marx, Karl. 1977. *Capital: A Critique of Political Economy*. Vol. 1, trans. by Ben Fowkes. Orig. published 1887. New York: Vintage.

Mukhia, Harbans. 1981. "Was There Feudalism in Indian History?" *Journal of Peasant Studies* 8, 3: 273–310.

Palan, R. 1992. "The European Miracle of Capital Accumulation." *Political Geography* 9, 4: 401–6.

Palat, Ravi Arvind. 1988. "From World-Empire to World-Economy: Southeastern India and the Emergence of the Indian Ocean World-Economy (1350–1650).'' Ph.D. diss., State University of New York at Binghamton.

———. (1991). "Symbiotic Sisters: Bay of Bengal Ports in the Indian Ocean World-Economy." In *Cities in the World-System*, edited by R. Kasaba, 17–40. Westport, Conn.: Greenwood Press.

Palat, Ravi Arvind, and Immanuel Wallerstein. Forthcoming. "Of What World-System Was Pre-1500 'India' a Part?" In *Merchants, Companies, and Trade*, edited by S. Chaudhuri and M. Morineau.

Perkins, Dwight H. 1969. *Agricultural Development in China, 1368–1968*. Chicago: Aldine.

Perlin, Frank. 1983. "Proto-Industrialization and Pre-Colonial South Asia." *Past & Present* 98: 30–95.

Polanyi, Karl, Conrad M. Arensberg, and Harry W. Pearson, eds. 1957. *Trade And Markets in Early Empires: Economies in History and Theory*. New York: The Free Press.

Ramaswami, Vijaya. 1985. *Textiles and Weavers in Medieval South India*. Delhi: Oxford University Press.

Rawski, Evelyn S. 1972. *Agricultural Change and the Peasant Economy of South China*. Cambridge, Mass.: Harvard University Press.

Reid, Anthony J. S. 1980. "The Structure of Cities in Southeast Asia: Fifteenth to Seventeenth Centuries." *Journal of Southeast Asian Studies* 11, 2: 235–50.

———. 1984. "The Pre-Colonial Economy of Indonesia." *Bulletin of Indonesian Economic Studies* 20, 2: 151–67.

———. 1990. "An 'Age of Commerce' in Southeast Asian History." *Modern Asian Studies* 24, 1: 1–30.

Richards, John F. 1990. "The Seventeenth-Century Crisis in South Asia." *Modern Asian Studies* 24, 4: 625–38.

Rostow, W. W. 1960. *The Stages of Economic Growth: A Non-Communist Manifesto*. Cambridge, England: Cambridge University Press.

Shiba, Yoshinobu. 1970. *Commerce and Society in Sung China*, trans. by M. Elvin. Ann Arbor: University of Michigan.

Sigaut, François. 1988. "A Method for Identifying Grain Storage Techniques and Its Application for European Agricultural History." *Tools & Tillage* 6, 1: 3–32.

Singh, Chetan. 1988. "Centre and Periphery in the Mughal State: The Case of Seventeenth-Century Panjab." *Modern Asian Studies* 22, 2: 299–318.

Skinner, G. William. 1974. *Marketing and Social Structure in Rural China*. Ann Arbor, Mich.: Association of Asian Studies.

———. 1985. "The Structure of Chinese History." *Journal of Asian Studies* 44, 2: 271–92.

Smith, Thomas C. 1980. *The Agrarian Origins of Modern Japan*. Stanford, Calif.: Stanford University Press.

———. 1988. *Native Sources of Japanese Industrialization, 1750–1920*. Berkeley: University of California Press.

Takaya, Y. 1975. "An Ecological Interpretation of Thai History." *Journal of Southeast Asian Studies* 6, 2: 190–95.

———. 1977. "Rice Growing Societies of Asia: An Ecological Approach." *South East Asian Studies* (Kyoto Daigaku Tonan Ajia Kenkyu) 15, 3: 442–51.

Tilly, Charles. 1990. *Coercion, Capital and European States,* AD *990–1990.* Oxford: Basil Blackwell.

Wallerstein, Immanuel. 1980. *The Modern World-System.* Vol. 2, *Mercantilism and the Consolidation of the European World-Economy, 1600–1750.* New York: Academic Press.

———. 1991. "World System versus World-Systems: A Critique." *Critique of Anthropology* 11, 2: 189–94.

———. 1992. "The West, Capitalism, and the Modern World-System." *Review* 15, 4: 561–619.

Washbrook, David A. 1988. "Progress and Problems: South Asian Economic and Social History, c. 1720–1860." *Modern Asian Studies* 22, 1: 1988.

Watabe, Y. 1978. "The Development of Rice Cultivation." In *Thailand: A Rice-Growing Society,* edited by Y. Ishii, trans. by P. Hawkes and S. Hawkes, 3–14. Honolulu: University of Hawaii Press.

Yamamura, K. 1979. "Pre-Industrial Land Holding Patterns in Japan and England." In *Japan: A Comparative View,* edited by A. M. Craig, 276–333. Princeton, N.J.: Princeton University Press.

5

Fatal Conjuncture: The Decline and Fall of the Modern Agrarian Order during the Bretton Woods Era

Resat Kasaba and Faruk Tabak

Changes in the agricultural landscape of the world-economy have of late received considerable attention as a result of the steady decline in the periphery's share in global agricultural trade (and production). One widely accepted interpretation attributes this decline to the effectiveness of the postwar food order that was set up under the aegis of U.S. imperial rule.[1] Stated briefly, this literature argues that the period spanning the Bretton Woods era nurtured and effectively administered a food order that has eventually altered the basic structures of world agriculture at the expense of the periphery.

We feel that the high "moment" of U.S. hegemony in the Bretton Woods era provides too short a time span to analyze adequately the changes in global patterns of agricultural production and trade. These changes flow from longer term trends and, as such, they need to be studied from a different historical vantage point. In what follows, we will claim that, when viewed from a long-term perspective, the Bretton Woods era appears to be of signal import but not because it involved the creation of a new food order undermining agricultural production in the periphery, and thereby extending the sphere of operation of American (or core) capital. Rather, the real importance of the postwar era lies in the more or less simultaneous completion of three overlapping and complementary cycles of different lengths. It was through the working of these cycles that world agricultural production became less a peripheral and more a "core-like" economic activity.

To start with, the Bretton Woods order signaled the end of the secular agricultural upswing which had started circa 1750. The commencement of a period of contraction in 1950–1970 is eloquently attested to by the ending at the zenith of U.S. hegemony of the inverse relationship between the prices of agricultural and manufactured goods which had long characterized world-economic fluctuations.

Roughly around the end of the Korean War, the volume of manufactured goods in world trade which, for nearly two centuries, had moved in tandem with the prices of agricultural and raw materials started to move independently of them. The reasons for the decoupling of these two sets of figures since 1952 were the gradual relocation of agricultural production in the core zones, and the ongoing increase in the share of intercore flows in world farm trade. The decline in the volume of agricultural trade is likely to continue into the next century in consonance with the demographic slowdown in the core zones. It is highly probable that the inflection of the postwar era represents the onset of a long-term agrarian contraction much like those that occurred in 1350–1450 and 1650–1750.

Second, the opening up of temperate settlements, especially of the New World, to full exploitation precipitated the completion of the life cycle of the Columbian exchange, which had started in the sixteenth century. The tight interconnection among the movement of prices of basic bread grains—the holy trio of wheat, maize, and rice—that resulted from the United States' pivotal position as the major exporter of all three crops is of paramount importance in charting the full consequences of Pax Americana. To be more precise, the significant role the United States has assumed since the 1950s in world rice markets has led to the transmission of the overall downward movement in world cereal prices to riziculture as well. As a result, the general downward pressure put on the prices of the world's basic bread grains has forced producers of these grains in the periphery to search for alternative cash crops, thereby extending further the share of core zones in the cereal trade.

Third, it was only in the Bretton Woods era that the axial division of labor, framed by the decline of Pax Britannica, was brought to full completion by Pax Americana. The decline of British hegemony from the 1870s on made possible the growing role of newly settled temperate settlements in world wheat trade, at the expense of the principal producers who had provisioned most of the European core during the mid-Victorian boom. These producers had been mostly, if not exclusively, of the periphery. Along with this decline in the periphery's share in world grain trade, the trade of temperate products have *grosso modo* become the almost sole preserve of the core zones, especially with the rise of Europe as a net exporter of food. In this, the establishment of a multilateral payments system in 1958–1959, which replaced the existing bipolar payments system, played a cardinal role in dismantling the last vestige of Pax Britannica. The widespread adoption of import-substitution programs by peripheral states has also contributed to the decline of agricultural trade in the periphery.

Despite the obvious differences in their historical span and breadth, these three developments have acted concurrently and have redefined the basic contours of world agriculture, whereby agriculture became a more corelike economic activity. Within this long-term world-scale restructuring, the munificent arrangements underwritten by the United States, important as they were, could not have been the main organizing force underlying the agrarian transformations of the postwar era; rather, the Bretton Woods era constitutes the *terminus ad quem* of these triadic cycles.

Historically, the limited effectiveness of the hegemonic arrangements put in place by the United States in shaping agrarian change was not an exception. In all three hegemonies of the capitalist world-economy, despite the hegemons' indisputable ability to singularly reshape and restore the global economic space, none has instituted a new food order. More typically, they revamped the order that came to embody their economic ascent and victory, an order that was not necessarily of their singular making.[2] Examined in this light, the period of hegemony appears as a moment—albeit critical, yet a "moment" nonetheless—in the longer process of agrarian change.

We will hence argue that it is not those fleeting moments of hegemonic supremacy but the long-term agricultural cycles and the drawn-out periods of hegemonic decline or ascent that should form and inform the contextual setting for depicting the patterns of long-term agrarian change. Conversely, we will claim that there are no substantial *historical* grounds for arguing that the institutional arrangements that were put in place by the hegemon played a constitutive role in recasting a new food order or regime.

The secular agricultural expansion that started in circa 1750, like those occurring between 1150 and 1350 and between 1450 and 1650,[3] was accompanied by a vigorous population rise in the core zones and by the incorporation of vast zones into the world-economy. Throughout this secular agricultural upswing, the relative volume of manufactured exports, mostly originating in the core, tended to rise in line with the rising incomes of primary producers, located mostly in peripheral, and later with the opening up of temperate settlements, increasingly in semiperipheral zones. During most of this period, when the terms of trade moved in favor of agricultural goods and raw materials, they usually led to an increase in the relative share of manufactures in world trade. Conversely, when terms of trade moved against agricultural goods, the relative share of manufactures fell.

The persistence of the secular cycle and its complementary price movements lasted until the 1938–1952 period (Rostow 1978, 91–99). In the aftermath of the Korean War, when prices of manufactures were rising relative to primary products, trade in manufactures did not decline but increased in tandem. This radical rupture in the cyclical movements of prices was caused by the growing concentration of farm trade in the core zones, by the rising share of petroleum in world trade, and by the increase in the volume of intercore trade in manufactured goods. Indeed, it was during the "golden age" of U.S. imperial order that agricultural trade as a proportion of all world trade fell from 33 percent to 17 percent (Van der Wee 1986, 107). These figures do not merely reflect short-term developments; they are the continuation of long-term trends that started at the turn of the twentieth century and involved the expansion of the share of manufactured goods in global merchandise flows. In fact, Pax Americana's "food order" helped somewhat conceal the termination of the secular cycle until the early 1970s, mainly by throwing noncommercial grain into circulation through its imperial redistribution programs. Since the possibilities of a shift toward either industrial crops or animal husbandry in the core zones are rather limited and since the deceleration in the rate of growth of the populations of these zones—populations now major consumers and producers of world grains—suggest that current trends will continue, we can expect that this secular downswing may be consolidated over time.[4]

Admittedly, the world grain trade might expand at times, for example, in the 1970s when the distribution of world liquidity favored peripheral states as well as agricultural producers in the core. But this was no more than a temporary reprieve as was demonstrated in the 1980s by the drastic decline in global demand precipitated by the policies of fiscal austerity imposed upon peripheral states. More important, since the onset of the present economic downturn, the declining ability of the world-economy's states to control global financial flows has rendered agriculture even more prone to the ongoing contraction in state-mediated transactions and flows which have been germane to its modus operandi.

Configurations of demographic decline and shrinking demand were the defining features of the period from 1350 to 1450 (following the Black Death and the Hundred Years' War) and of the period from 1650 to 1750 (following the Thirty Years' War) (Van der Woude 1992). To recapitulate, then, the Bretton Woods period witnessed and even secured the beginning of a secular agricultural contraction, not unlike those that took place at the end of the Middle Ages and of the seventeenth century. Enter the secular downswing.

The second world-historical process that came to a close in the Bretton Woods era was the Columbian exchange. Even a brief examination of the closing of this life cycle forces us to shift our focus beyond the narrow confines of the U.S. food regime. Historically speaking, the westward expansion of agricultural cultivation in the United States intimately intertwined during the mid-nineteenth century the destinies of two grain crops which had crossed the Atlantic Ocean in opposite directions during the sixteenth century: maize and wheat. By the end of the nineteenth century, maize, a veteran of the Columbian exchange, became a widely cultivated crop, harvested from the vast stretches of inner China to the deep corners of Africa (Latham 1981). In a similar fashion, wheat cultivation gained a foothold on both hemispheres of the American continent, to expand spectacularly during the latter half of the nineteenth century.

With the expansion of rice cultivation in the late nineteenth century, the concentration of the production of these three bread crops—the staple foods of the majority of the world's inhabitants—in one locale, that is, in the United States, facilitated the transmission of price fluctuations to the homelands of these crops, from Monsoon Asia to Latin America and Africa. The fall in wheat prices was, then, but one aspect of the restructuring of world agriculture. The downward pressure exerted by falling wheat prices upon those of maize and rice extended the stretch of the agrarian depression far beyond the Old World, to the distant corners of the globe, affecting grain producers high and low (Barker, Herdt, and Rose 1985). It is by gauging and mapping the interconnections and symbiotic relationships between the prices and the production of these basic grains—of what Fernand Braudel referred to as "the plants of civilization" (1981, 107), which together account for over two-thirds of the world's supply—that the ramifications of the agricultural transformations can be fully traced.

When viewed from the vantage point of the world-economy, the "wheat standard," as it were, precipitated a series of developments: The decline in the price of rice and maize during the 1930s, though less steep than that of wheat, engendered a reduction in the cost of reproduction of labor. This, in turn, fostered the dissemination and consolidation of plantation economies, most notably in Asia and Africa, and increased the specialization of the periphery in the cultivation and procurement of raw materials and tropical crops (Hanson 1980). Second, the fall in wheat prices facilitated the diffusion during the depression of maize and rice cultivation which both commanded relatively higher prices than wheat. But the overall downward trend imposed by wheat on other grain crops also helped deepen the commercialization of the nonplantation sectors in these regions. That rice could be, and at times was, substituted for by wheat is clearly evident in the countercyclical movements of their production patterns, both in the closing decades of the nineteenth century (Latham and Neal 1983) and in the interwar period (Yates 1959). Third, the fall in the price of maize in Latin America released former small farmers from their holdings and turned them into reserve labor to haciendas. When maize production for (European and Argentine) livestock industries started to expand in harmony with the spec– tacular expansion of wheat cultivation in the pampas at the turn of the century, the concentration of production of both crops in haciendalike units and the consequent separation of producers from land prepared the ground for the im- port-substitution policies of the interwar years (Mörner 1977). Even though a relatively small proportion of these crops was thrown into the global circuits of trade, the fall in their prices had a wider impact on a plethora of producers.

Depicting the fall in wheat prices, primarily through the downward pressure it exerted on wages in the core (and the role it played in fostering industriali- zation), without specifying its geographical scope, fails to capture the full array of changes characterizing this period. Judging the significance of grain trade by its sheer volume fails to capture these changes in all their diversity. Small though the share of exports was with respect to world production, its effects were far reaching because the end result was not necessarily growing commercialization of a certain crop, but growing integration between the markets and prices of bread grains, which facilitated the transmission of price fluctuations across regions and continents. Since the prices of the three basic grains have started to move in the same direction in the postwar period, it is no longer possible to substitute one for the other as was done, say, in Asia at the end of the nineteenth century and during the 1930s.[5]

Examining the position the United States occupies as a significant producer of the world's civilizational crops puts the gradual constitution of U.S. imperial rule into its proper historical perspective rather than attributing undue emphasis to its "golden age." The specificity of the U.S. hegemony within the realm of agriculture can then be attributed more to its success in completing the Columbian exchange in the post-1945 period than to the food aid and concessionary sales programs of the Bretton Woods era. Exit the Columbian exchange.

The third and the last cycle we will trace dates back to the period from 1873 to 1896, the period known as the Great Depression of the nineteenth century. Here, we perforce narrow our focus from the plants of civilization to temperate crops, more specifically to wheat. As a result of the economic structuring that took place during the demise of Pax Britannica, the temperate settlements became the main suppliers of grain to the core areas. Two developments, both part and parcel of Pax Americana, progressively reinforced the foundations of this division of labor over the course of the twentieth century: (1) the contraction of the periphery's share in world agricultural production and (2) the gradual concentration of agricultural production and trade in the core.

In the immediate aftermath of the mid-Victorian boom, the adoption of a silver or bimetallic standard by peripheral states as opposed to those of the temperate settlements, who adhered to the gold and sterling standard, enabled the former group to take advantage of the steady depreciation of their currencies (Nugent 1973). As the gold price of silver declined, the agricultural exports of the peripheral zones actually increased momentarily during the late nineteenth century. These trends were reinforced by the establishment of currency blocs which overshadowed, to an extent, the periphery's weakening position within the global division of labor during the interwar period. To some extent, the number of peripheral countries depending on one (nongrain) product for more than half of their export proceedings increased in the interwar era (Yates 1959).

However, the initial reprieve the agricultural sector received from these measures and from the import-substitution strategies (such as the overvaluation and nonconvertibility of currencies) proved to be temporary. In the former European colonies (mostly in Asia and Africa), these policies found a protected environment with the breakup of the unity of the world-economic space during the interwar era, and, more important, in the limited sway of American corporate capital in the 1950s and 1960s. In these locales, the agricultural sector provided the industrial sector with the necessary labor force, with food for soaring urban populations, and with foreign exchange (via exports). But overall, outmigration from rural areas and, more important, changes in crop mixes at the expense of basic grains—the cultivation and flow to urban areas of which were under the scrutiny of statal agencies—engendered a decline in cereal cultivation.

In Latin America, which was within U.S. imperial dominion from early on, import substitution was financed partly by U.S. corporate capital. It should be noted here that, at this stage, the U.S.-based capital had to confine its operations to the dollar zone (or the Western Hemisphere) because of the difficulties of establishing a multilateral payments system. In most of Latin America, the force and magnitude of industrialization was such that the agricultural sector was quickly dwarfed. In the end, by design or by default, the result was the contraction of the share of the periphery in world grain trade. Therefore, the corresponding contraction in the share of the peripheral zone was deepened by widespread subscription to the import-substitution policies of the inter- and postwar eras—policies designed, in part, to arrest the above-mentioned contraction but ironically served to further it because of the part agriculture was expected to play in the process of industrialization.[6]

Two developments temporarily obscured the scale of the contraction of agricultural production in the periphery during the Bretton Woods era. The first was the food aid program and the concessional sales where were seen as the pivotal components of the Pax Americana. After reaching as high as 36 percent of world exports in the mid-1960s, their share declined to about 17 percent in the early 1970s. Right around that time, the flush of liquidity in world financial markets enabled the peripheral states to finance their imports of agricultural goods, thereby obscuring further the continuing contraction of the agricultural sector in these years.

The debt crisis of the 1980s and the subsequent retrenchment of capital at core locales eventually helped reveal the magnitude of the periphery's shrinking share in world agricultural production. Drastic reduction of imports, caused by the drying up of capital flows, was instrumental in crystallizing this secular trend during the 1980s (Buttel 1989). Moreover, the almost geometric rate at which the populations of the periphery have been growing since the 1950s has further dwarfed the share of these regions in world agricultural trade. Cereal imports by these regions, which stood at 6 million tons between 1948 and 1952, are estimated to reach from 100 million to 200 million tons by the end of the century (Bairoch 1988), while Europe and continental North America together accounted for over 80 percent of all "traded" cereals. In brief, a process the foundations of which were laid by the decline of British hegemony has been furthered and completed by U.S. hegemony.

Another strand of continuity that links the interwar era with the Bretton Woods period is the growing share of inter- and intracore grain trade in the total world trade in grain since the turn of this century (Bacon and Schloemer 1940; Van der Wee 1986). This suggests that, as far as trade in agricultural goods was concerned, the high point of U.S. hegemony does not seem to have involved a break with the interwar period. Furthermore, despite the significance attached to food aid programs in expanding American exports to the periphery, over three quarters of world trade continued to take place within and among the core and semiperipheral zones during the postwar period.

In other words, the patterns of agricultural production and trade that are associated with the periods of U.S. hegemony came about not through a set of distinct institutions—hence a food order—that were set up in the postwar era, but through a series of transformations that are traced back to the nineteenth century. The ascendance of temperate settlements as the principal sites of grain production and the systemic implications of this spatial patterning on the "multilateral trade imbalances" was an integral part of the imperial rivalry that led to the gradual erosion of Pax Britannica (Frank 1976). These settlements received a big boost, not only in terms of labor and capital inflows, but also from the Imperial Preference System—both of which placed them at the very heart of the U.K.–centered world-economy. This pivotal position allowed "overseas" producers to establish firm command over world agricultural production and trade at the expense of the peripheral producers, who were not equipped with similar cushioning mechanisms.

The proliferation of protectionist measures in the European core was further supplemented by the formation of "currency areas" and "trading blocs," both of which gave the agricultural sector a precarious reprieve during the interwar era. Gradual though it might be, the increase in the share of grain-importing European countries in world agricultural production was symptomatic of the developments to follow. Whereas production by European exporters increased from 26 percent of the world total from 1919 to 1924 to 35 percent from 1934 to 1939, the share of overseas exporters fell from 42 percent to twenty-eight percent during the same period (Malenbaum 1953, 239).

First the "conservative thirties" and later the absence of a global multilateral payments system, as well as the incovertibility of sterling, precluded the translation of the agrarian order in the making during the interwar period into a global order. Inclusion into the Sterling Area, which enveloped most of the temperate settlements, tethered the volume of exports originating there to the ebbs and flows of a gradually weakening pound sterling. However, a series of developments rendered inclusion in the Sterling Area not as burdensome: The devaluation of sterling in 1949, the outbreak of the Korean War in 1950, and, in general, the terms of trade favorable to agriculture in the period from 1936 to 1952 helped alleviate the burden of being a part of the Sterling Area. It was in 1951–1952 that the world grain trade reached the record level of almost one billion bushels, only slightly higher than that reached in 1928–1929, 950 million bushels.

As a result of their favorable balance of payments, many Commonwealth nations and others were able to accumulate large sterling balances during and after World War II. Although this strengthened the reserve position of the members of the Sterling Area, it did not resolve the underlying problems that stemmed from the fact that two members of the temperate settlements were in the dollar zone; one, in fact, constituted the epicenter of it. (Of all the countries and territories of the British empire, only Canada did not link its currency to the pound.) The absence of a unitary multilateral payments system during the late 1940s and the 1950s found expression in the fragmented structure of the world grain market: Argentina catered to the OECD countries, Australia to Asia, and Canada and the United States to the world, until of course Europe reestablished its position.

The results of the dissolution in 1958 and 1959 of the European Payments Union (EPU) and the resumption of a multilateral payments system were immediately visible.[7] Between 1958 and 1965, intracommunity as well as extra-community trade in cereals soared; the former tripled in volume and the latter increased significantly. France increased its cereal exports by about 500 percent, and Britain resurfaced as the biggest importer of cereals. Overall, the total world cereal production rose by about one-third between 1961 and 1965 and 1971 and 1973. Consequently, cereal production per capita at the core rose about 25 percent and very little in the periphery. To put it differently, the trends that had started to establish the core zones as major sites of cereal production found their full expression only in the 1960s, but the nascent order was not the creation of the imperial arrangements of the postwar period. In addition to the ongoing shifts in sites of production, the protectionist policies of the core states during the interwar era and the discrimination Europe and Japan were accorded by the United States against its exports during the reconstruction period (and afterwards) reinforced the division of labor in agricultural production.

Accordingly then, in the crystallization of this new agrarian order, we think that the dissolution in the late 1950s of the EPU and the establishment of the full convertibility of the sterling played a more crucial role than acknowledged in the literature. It was with these developments that the relocation of agriculture in the core, which had been triggered by the decline of British hegemony, was finally completed. Exit Pax Britannica.

Within this frame of inquiry, U.S. food aid played a supporting part until the resumption of exports from the European core was resumed. Food aid programs were no more than temporary measures designed to reproduce the interwar agricultural programs. In some ways, they were similar to the wartime loans to the Allied powers and the earlier lending operations of the 1920s. It was the "easy" credit which, in both instances, played a key role in the export of grains. Price support programs before and after the war domestically rendered the perpetuation of the agri-food system relatively easier for the U.S. state.

Hence, though the undoing of the dual payments system, which helped resume world grain trade, was related to the consolidation of U.S. hegemony, the dwindling of American food aid was neither coterminous nor synonymous with the demise of the U.S. role. The final vestiges of British imperium disappeared with the general acceptance of the gold-exchange standard and the establishment of the dollar as a reserve currency. Along with these developments, the geographical sweep of U.S. imperial order encompassed most parts of Africa and Asia. And with the establishment of Pax Americana proper, food aid became a thing of the past. World-economy-wide restructuring of agricultural production, which prompted increasing grain exports from Europe, rendered food aid incompatible with the U.S. imperial order proper. Hence, the reconstruction of Europe and the adherence to import-substitution policies, facilitating the restoration of the interstate system as well as the transnational expansion of capital under the aegis of U.S. imperial rule, accelerated the underlying tendencies that had been shaping world agriculture since 1873.

The limited reach of the United States in reshaping world agricultural production is not a unique feature of Pax Americana. Historically speaking, in the case of Pax Neerlandica, the pattern of specialization furthering Amsterdam's mercantile control over the vibrant Baltic grain trade during the moment of Dutch hegemony failed to survive into the latter half of the seventeenth century (De Vries 1974). What followed was not a furthering of the Baltic grain trade—the Soul of all Commerce—which had propelled Amsterdam to her pivotal role in redistributing the grain originating in the east of Elbe to northwestern and southern Europe. On the contrary, after 1650, this *staplemarket* gradually declined, and grain production shifted from the periphery to the core and semiperiphery (Wallerstein 1980, 166–67).[8]

The agrarian patterning of Pax Britannica was no different. The global division of labor sanctioned and strengthened by the repeal of the Corn Laws in 1846 did not last into the last quarter of the century. It was during the mid-Victorian boom that the abolition of the Corn Laws fueled grain production in the Danube basin, New Russia, the middle and lower Volga, and India. Yet, it was not these locales but the temperate settlements which, with the start of the Great Depression of the nineteenth century, came to establish their command over world grain production and trade, thereby dramatically altering the geography of agricultural production which had characterized the heyday of British imperial rule.

The fact that imperial mechanisms available to the hegemon were ineffective in setting a framework for a new order is a testimony to the observation that these are primarily designed to restore and expand the already existing relations, not to create new institutional arrangements. The Dutch profited from an existing grain trade—grain from the Vistula—centered around the Baltic Sea, a product of the expansionary sixteenth century. To be sure, its dominance in this trade, established already in the 1580s, was further consolidated by the strength of its shipbuilding industry and the rebuilding of the Amsterdam Bourse immediately after the large-scale shipping of grain to the Mediterranean shores began in 1590–1591 (Braudel, 1981, 208). The end of the Thirty Years' War and the economic climate ensuing the Treaty of Westphalia allowed the expansion of the agrarian base in the core and semiperipheral zones. This process was considerably facilitated by the introduction in Europe of New World crops, such as potatoes and maize, and American rice, which brought a further shift in the locus of cereal production (Le Roy Ladurie and Goy 1982). Conversely, production in the peripheral areas of the seventeenth century, that is, Eastern Europe and Hispanic America, started to cater primarily to regional markets, as partial withdrawal from production for the market got under way (Wallerstein 1980, 131).

The British hegemony, on the other hand, spanned the period of incorporation of virtually all parts of the globe, a process framing the establishment and extension of its command over global economic flows in the course of the eighteenth century. Again, it was the ability of the British to fashion an "informal" empire—that is, to establish the unity of the global economic space—which allowed them to reproduce the existing division of labor. The abolition of the Corn Laws reinforced existing patterns of grain flows, now amplified in volume, but did not fashion a new network of production and trade. The imperial machinery at the hegemon's command was utilized to sustain the already existing arrangements. With the intensification of imperial rivalry in the last quarter of the century, the economic space created by the pax was, as to be expected, badly fractured both by the "scramble for colonies" and by the transformation of Britain's informal empire. The transformation of this informal empire into, in effect, a formal empire enveloping the vibrant temperate settlements enclosed capital and labor flows within and among the British isles, continental Europe, and the temperate settlements. The creation of this integrated network, in turn, undermined the position of the former sites of grain production.

As can be surmised, in each instance, the order that laid the groundwork for the hegemon's rise was undone by the end of the hegemonic "moment," if not earlier. After the middle of the seventeenth century, Baltic ports supplied the northern Netherlands with less grain than in earlier decades, and the voluminous grain trade decreased steeply. By the first half of the eighteenth century, the average quantity of grain passing through the Sound had declined by about one half as compared with the first half of the seventeenth (Faber 1966, 118). Likewise, the informal empire, which allowed the United Kingdom to further its economic lead, was transformed into a formal empire, consolidating the rise of temperate settlements at the expense of the former granaries of the empire. The United States, as a member of the temperate settlements, benefitted from the order established during the demise of Pax Britannica, yet it was during its own hegemony that the system that had served it so well was eventually undone.

That periods of hegemonic decline were sites of long-term reorganization of world agriculture is also manifested in the historical breadth of the ensuing social arrangements. To wit, the demise of the food order shaped by Dutch hegemony fostered the rise of *gutsherrschaft*-like structures to the east of Elbe and haciendas in Hispanic America. Both forms of organization of production lasted well into the latter half of the eighteenth century (cf. Goldstone 1991). In a similar fashion, the period following British hegemony witnessed the rise and consolidation of family farming in select locales, setting the much touted Chayanovian dynamics in motion, in a revamped form of course. Again, the organization of production along these lines lasted well into the second half of this century.

In the light of this historical patterning common to the hegemons, it becomes clear that the solidity and unity attributed to periods of hegemonic maturity, due to the brevity of these periods, tends to exaggerate the constitutive role the hegemons play in shaping the global agrarian arena. The food order theory is no exception. In taking stock of the transformations in world agriculture since 1945, it operates through and privileges two closely related and overlapping assumptions. The first assumption is that the U.S. imperial order, as codified at Bretton Woods, played a pivotal role in the transformation of world agriculture. The second assumption derives from the centrality attributed, à la Malen–baum (1953), to the state of "disequilibrium" between world supplies and demands for wheat. This state of disequilibrium, as one of the basic attributes of the interwar period, is said to have been redressed, in the years following the war's end, by the redistributive arrangements of the U.S. imperial order, inter alia, the Marshall Plan, the reconstruction of Japan, the Food for Peace Program, and the infamous PL 480. These policies in unison account for the part played by the United States in reshaping the structure and geography of world agriculture. Hence, a radical break with the interwar period is inscribed in the codification of U.S. imperial role and the installation of its redistributive mechanisms. Both assumptions, we have argued, need to be reevaluated in the light of the historical trajectory charted above in rather broad strokes.

Seen from this historical altitude, then, the world-historical developments attributed to the moment of American hegemony capture only a moment of world-scale agrarian structuring. Analyses that dwell upon short-run evaluations of these wide-ranging developments point to a directional change in the provenance of agricultural exports only in the 1970s and then attribute this contraction largely to the success of American food aid programs. When placed in its historical context, it becomes evident that the order in place now has long been in the making and was essentially framed by the aforementioned family of cycles. It can neither be attributed to the policies and developments of a decade or two, nor solely to the institutions of U.S. hegemony. While the particular role the United States will play in the coming "post-Fordist," "post–Cold War," and "postmodern" world is far from clear, there is little doubt that the patterns of agricultural production and specialization in the making will continue to dominate the world-economy for some time to come.

NOTES

The authors acknowledge the assistance of Philip McMichael in revising this chapter.

1. Here we are referring to a series of articles written by Harriet Friedmann and Philip McMichael. See Friedmann (1982, 1993); McMichael (1993); and Friedmann and McMichael (1989).

2. On hegemony, see Research Working Group on Cyclical Rhythms & Secular Trends (1979); Wallerstein (1984); Arrighi (1990); and Hopkins (1990).

3. The secular agricultural expansion extending from 1750 to circa 1950–1970 is different from Rondo Cameron's logistics, although they both cover the same chronological time span. Cf. Cameron (1978). On secular agricultural cycles, see Abel (1978); Slicher van Bath (1963); and Le Roy Ladurie and Goy (1982).

4. In the analysis of secular cycles, alongside demography, the acreage under cultivation is also taken into account. Yet for problems arising in calculating agricultural land

under arable cultivation in view of wide-ranging land-use patterns, prevalent especially in the periphery, see Boserup (1974).

5. For the price movements during the great depression of the nineteenth century, see Latham and Neal (1983); for that of the interwar period, see Yates (1959). For a documentation of recent trends, see Tyers and Anderson (1992).

6. On the effectivity of policies the core states were forced by the peripheral states to initiate (e.g., Stabex) in order to alter the contraction in the periphery's share in world agricultural trade, see Bessis (1985); also the incisive evaluation by Emmanuel (1976).

7. For a detailed account of the transformation of the postwar payments system, see Van der Wee (1986) and Hirsch and Oppenheimer (1977).

8. Whether it was the Dutch predominance in bulk trade in itself, or the addition to this of "rich trades," as argued by Israel (1989), that led to hegemony is of no consequence within the context of our argument.

REFERENCES

Abel, Wilhelm. 1978. *Agricultural Fluctuations in Europe from the Thirteenth Century to the Twentieth Century.* 3d rev. ed. New York: St. Martin's Press.

Arrighi, Giovanni. 1990. "The Three Hegemonies of Historical Capitalism." *Review* 13, 3: 365–408.

Bacon, L. B., and F. C. Schloemer. 1940. *World Trade in Agricultural Goods.* Rome: International Institute of Agriculture.

Bairoch, Paul. 1988. *Cities and Economic Development.* Chicago: Chicago University Press.

Barker, R., R. W. Herdt, and B. Rose. 1985. *The Rice Economy of Asia.* Washington, D.C.: Resources for the Future.

Bessis, Sophie. 1985. *L'Arme alimantaire.* Paris: Maspero.

Boserup, Ester. 1974. "Food Supply & Population in Developing Countries." In *Agricultural Policy in Developing Countries*, edited by N. Islam, 164–76. New York: St. Martin's Press.

Braudel, Fernand. 1981. *Civilization & Capitalism, 15th–18th Century.* Vol. 1, *The Structures of Everyday Life.* New York: Harper & Row.

———. 1984. *Civilization & Capitalism, 15th–18th Century.* Vol. 3, *The Perspective of the World.* New York: Harper & Row.

Buttel, Frederick H. 1989. "The US Farm Crisis and the Restructuring of American Agriculture: Domestic and International Dimensions." In *The International Farm Crisis*, edited by D. Goodman and M. Redclift, 46–83. London: MacMillan.

Cameron, Rondo. 1978. "The Logistics of European Economic Growth: A Note on Historical Periodization." *Journal of European Economic History* 2, 1: 145–48.

De Vries, Jan. 1974. *The Dutch Rural Economy in the Golden Age, 1500–1700.* New Haven, Conn.: Yale University Press.

Emmanuel, Arghiri. 1976. "La 'Stabilization': Alibi de l'exploitation internationale." *Revue Tiers-Monde* 17, 66: 257–76.

Faber, J. A. 1966. "The Decline of the Baltic Grain-Trade in the Second Half of the 17th Century." *Acta Historiae Neerlandica* 1: 108–31.

Fontana Economic History of Europe, edited by C. M. Cipolla, Vol. 2, *The Twentieth Century*, 603–97. London: Fontana.

Frank, André G. 1976. "Multilateral Trade Imbalances and Uneven Economic Development." *Journal of European Economic History* 5, 2: 407–38.

Friedmann, Harriet. 1982. "The Political Economy of Food: The Rise and Fall of the Postwar International Food Order." *American Journal of Sociology* 88 (Supplement): 248–86.

―――. 1993. "The Political Economy of Food." *New Left Review* (January/February) 197: 29–57.

Friedmann, Harriet, and Philip McMichael. 1989. "Agriculture and the State System." *Sociologia Ruralis* 29, 2: 93–117.

Goldstone, Jack A. 1991. *Revolution and Rebellion in the Early Modern World.* Berkeley: University of California Press.

Hanson, John R. 1980. *Trade in Transition: Exports from the Third World, 1840–1900.* New York: Academic Press.

Hirsch, Fred, and P. Oppenheimer. 1977. "The Trial of Managed Money: 1920–1970." In *The Fontana Economic History of Europe*, edited by Carlo Cippola, vol. V, part 2, 613–97. London: Harvester.

Hopkins, Terence K. 1990. "Note on the Concept of Hegemony." *Review* 13, 3: 409–11.

Israel, Jonathan I. 1989. *Dutch Primacy in World Trade, 1585–1740.* Oxford, England: Clarendon Press.

Latham, A.J.H. 1981. *The Depression and the Developing World, 1914–1939.* Totowa, N.J.: Barnes & Noble Books.

Latham, A.J.H., and Larry Neal. 1983. "The International Market in Rice and Wheat, 1868–1914." *Economic History Review* (2d series), 36, 2: 260–80.

Le Roy Ladurie, E. Goy, and J. Goy. 1982. *Agricultural Fluctuations.* Cambridge, England: Cambridge University Press.

Lewis, Arthur, ed. 1970. *Tropical Development, 1880–1913.* London: George Allen & Unwin.

McMichael, Philip. 1993. "Agro-Food Restructuring in the Pacific Rim." In *Pacific-Asia and the Future of the World-System*, edited by Ravi A. Palat, 103–16. Westport, Conn.: Greenwood Press.

Malenbaum, Wilfred. 1953. *The World Wheat Economy, 1885–1939.* Cambridge, Mass.: Harvard University Press.

Mörner, Magnus. 1977. "Latin American 'Landlords' and 'Peasants' and the Outer World during the National Period." In *Labour and Land in Latin America*, edited by K. Duncan and I. Rutledge, 455–82. Cambridge, England: Cambridge University Press.

Nugent, Jeffrey. 1973. "Exchange-Rate Movements and Economic Development in the late Nineteenth Century." *Journal of Political Economy* 81: 1110–35.

Research Working Group on Cyclical Rhythms and Secular Trends. 1979. "Cyclical Rhythms & Secular Trends of the Capitalist World-Economy." *Review* 2, 4: 483–500.

Rostow, Walt W. 1978. *The World Economy: History and Prospect.* Austin: University of Texas Press.

Slicher van Bath, B. H. 1963. *The Agrarian History of Western Europe, A.D. 500–1850.* London: Edward Arnold.

Tyers, Rod, and Kym Anderson. 1992. *Disarray in World Food Markets.* Cambridge, England: Cambridge University Press.

Van der Wee, Herman. 1986. *Prosperity & Upheaval: The World Economy, 1945–1980.* Berkeley: University of California Press.

Van der Woude, Ad M. (1992). "The Future of West European Agriculture: An Exercise in Applied History." *Review: A Journal of the Fernand Braudel Center* 15, 2: 243–56.

Wallerstein, Immanuel. 1980. *The Modern World-System.* Vol. 2, *Mercantilism & the Consolidation of the European World-Economy 1600–1750.* New York: Academic Press.

———. 1984. "The Three Instances of Hegemony in the History of the Capitalist World-Economy." *International Journal of Comparative Sociology* 24, 1/2: 100–108.

Yates, P. Lamartine. 1959. *Forty Years of Foreign Trade.* New York: MacMillan.

PART III

Contemporary Agro-Food Complexes

6

Canadian Misfortunes and Filipino Fortunes: The Invention of Seaweed Mariculture and the Geographical Reorganization of Seaweed Production

Lanfranco Blanchetti-Revelli

On May 10, 1976, the president of the Canadian subsidiary of Marine Colloids Inc. (MCI), the biggest processor of carrageenan in the world, wrote the following advice to the Prince Edward Island Market Development Center:

I am going to start off by telling you that in 1976 we will be taking the smallest amount of Irish Moss from Canada that we ever have since I joined the company in 1967. This is due to a variety of reasons, the two main ones being that there is a surplus of seaweed on the world market at this time and that the cheapest ones are getting acceptance. (Cited in Anderson, Frenette, and Webster 1977)

The news was bad for the hundreds of people who made a living from gathering moss off the Atlantic coast of Canada. In fact, since then, that region's moss production, which had grown constantly for twenty years, began to fall on a regular basis season after season.

On the other side of the globe, however, among Filipinos in Cebu and in the Sulu Archipelago, the same news would have been taken with satisfaction—by hundreds of people who had just begun to exploit seaweed a few years before. The reason for the seaweed surplus was in fact due to a growing production in the Philippines.

By focusing on various events—a technological innovation, a market crisis, a strategic move on the side of corporate business—this chapter will examine how, as the demand for carrageenan and its raw materials increased, a world relocation of the major seaweed-producing areas took place. The following pages describe how interests located in the metropolitan core of the world-economy can follow strategies that cause alternate fortunes for different populations located in various parts of the periphery. Without denying the general pattern of dependency entailed by the structure of the modern world-economy, the material presented seems to suggest that dependency occurs in different degrees and modalities according to specific local conditions in distinct points of the periphery (Cardoso 1977; Mintz 1977). Simultaneously, however, this material suggests also that the degree of dependency is relative to factors specific to the commodity under examination: for example, the commodity's positioning in broader world economic dynamics, its productive or trade requirements, its industrial maturity, and last, but not least, corporate strategies.

CARRAGEENAN AND SEAWEED

Carrageenan is a commodity little known to the general public. Unlike some commodities, which are used and consumed everywhere in the world and have properties and shapes known directly by the consumer, carrageenan is, as a commodity, a sort of "ghost." Not only do those who eat it not know where it comes from (a slimy sea plant they would avoid with disgust during a vacation at the beach), not only do they not have any idea of its refined form (a white, yellowish powder) but also they often do not even know that they are eating it when they are.

Carrageenan, however, is well known for its extreme efficiency and adaptability as a thickener, as an emulsifier, and as a binder to those concoction engineers who are behind the prepackaged foods eaten by millions of consumers everywhere in this "post-cooking" "post-everything" reconstituted world. In the past twenty years, carrageenan has been used in increasing quantities by alimentary and nonalimentary industries in an ever expanding array of applications. Nowadays carrageenan is used in milk and dairy products, beer manufacturing, sweets, compound meats, and fat-free products, for example, the greatly publicized McDonalds' McLean hamburger.

When one speaks of the efficiency of a product in the context of modern industry one does not refer only to the technical properties of a substance but more precisely to a subtle balance between properties and costs. Even when a particular substance appears perfect from the technical point of view, its industrial efficiency can be measured only in terms of the technical properties and the cost of other substances that can be used as substitutes. After having been used for centuries as raw dried seaweed in Ireland, Scotland, and northern France in the confection of milk puddings, carrageenan was first synthesized at the end of last century. For several decades, its use was limited to a few specific applications. Its cost and the irregularity of supply of raw materials—a red seaweed called *Chondrus crispus* but better known as Irish moss—limited its utilization to those applications that were so specialized that cost was irrelevant.

Industrial carrageenan refining began in the United States, where the use of Irish moss, introduced by Irish immigrants, was spurred by vast Irish moss beds off the coasts of New England and especially the Canadian Maritime Provinces. During World War II, carrageenan grew in importance as a partial substitute for agar[1] which was the near monopoly in Japan. After the war, thanks to the availability of raw materials, the discovery of new extractive methods, and new applications, the demand for the product expanded enormously. While Canada became the world's biggest producer of carrageenophytes, two processors in the United States, MCI and Stauffer Chemicals; two in Denmark, Litex and Kopenhagen Pectine Factory (KPF); and one in France, Peyrefitte-Auby—the result of new initiatives and mergers between small operators—achieved control of the market. Thus, the emerging structure of carrageenan markets took the form of an oligopsony and entailed a clear geographical separation of raw materials producers and processors. In fact, as Danish and French sources were insufficient for the needs of local manufacturers, these began increasingly to depend on Canadian raw materials.

CANADIAN SEAWEED AND CANADIAN HARVESTERS

Between 1948 and 1974, the year of highest production, the total landings of seaweed in the Maritimes went from 2,582 wet metric tons (1,314 dry) to 50,400 wet metric tons (12,600 dry) (Pringle and Mathieson 1987). Initially, Canadian moss was gathered by farmers and settlers who went to the shore after heavy storms to load carts and trucks with the product tossed on the beach by wind. Farmers also used hand rakes to rip off the seaweed from the beds closer to land. By the late 1940s, fishermen joined the enterprise and began to exploit more distant beds by dragging big rakes with their boats. Before the introduction of harvesting regulations, moss was harvested from April, when the ice breaks, onward. Later, with the establishment of regulations aimed at maintaining the reproductive potential of seaweed beds, mossing was allowed only between June and October.

In the Maritimes seaweed harvesting is practiced by settlers and fishermen in combination with other activities, farming, and especially lobster fishing. Second in importance only to this latter activity, mossing contributes significantly to the yearly income of the local populations and offers a substantial contribution to the economy of a region that has few resources and is plagued by a very high rate of unemployment. In the mid-1970s, seaweed harvesting and gathering was the mainstay for more than 750 households; another 1,000 to 1,500 households engaged in the activity as a secondary source of income (Anderson, Frenette, and Webster 1977). Still in the same period, the seaweed industry offered seasonal employment to from 200 to 300 persons who worked in the buying sector as driers, stevedores, and brokers.

Historically the political economy of the Maritime Provinces has always been characterized by highly unequal productive relations and by a great dependence on external financial sources. Since before the development of a seaweed economy, the lobster fisheries in the Maritimes were controlled by a few nonlocal corporations and dozens of small family-operated canneries. These operators had a tight grip on producers who went fishing on corporation-owned boats and were often so indebted to them that they rarely received cash in payment for their lobsters. Because the canneries could not resist the corporations' competition, whole villages became indebted to external operators. After the war, thanks to a more intensive financial intervention of the state, many fishermen were able to purchase boats and implements and, following the establishment of fishermen's cooperatives, managed to build a relative independence from the corporations.

At the beginning, seaweed harvesting was not as tight in terms of financial requirements as fishing. Gathering sea moss on the beach does not involve expensive implements. Seaweed exploitation, however, becomes more dependent on capital when it is carried out intensively. In such a case, it requires a boat, gasoline, and other devices, including winches, rigging, and racking equipment. Apparently, lobster fishing and mossing developed symbiotically. Initially, the income provided by gathering shore-tossed seaweed contributed to the lessening of the lobster fishermen's dependence on borrowing. Later, once the fishermen began to own their own boats, lobster fishing helped in expanding mossing to remote beds. This fact increased seaweed harvesting capabilities so dramatically that, once a stable market was established, the new activity became the standard substitutive activity during the lobsters' molting season.

But, if mossers were relatively independent in terms of control of means of production, they were still subjected to the buying policies of few local firms relying on foreign corporations for marketing outlets and, therefore, revenue. As great as it was in the first decades of mossing activities, the buyers' grip on producers was bound to grow even tighter in the mid-1960s. This occurred when MCI and the two Danish processors incorporated subsidiaries in Canada and opened buying stations in the Maritimes. The corporations' move followed a cost-cutting rationale and aimed at bypassing intermediate marketing structures. Soon the local buyers were either put out of business or absorbed into the corporate buying structure. Then, notwithstanding the rise of the world carrageenan demand caused by the expansion of its uses, the corporations used the power given to them by the oligopsonistic structure of the seaweed market to contain price escalation.

The corporations' arrival in the Maritimes brought about a sudden growth in local seaweed demand. In order to satisfy their procurement needs, the multinationals set mechanical drying facilities allowing the absorption of raw materials also with bad weather. The productive growth caused by this development was paralleled by a growth in mossers' incomes. This improvement, however, was offset by the fact that, in order to gather greater amounts of seaweed, the mossers had to undergo costly upgrades of their racking technology. Thus, while the corporations enforced their oligopsonistic power to contain price escalation, many dissatisfied mossers began to look for ways to counterbalance corporate policies. Aware of the high demand for their product, they began to ask for higher prices. Finally, in 1971, in order to give force to their requests, they went on strike.

Another form of resistance against the corporations was the establishment of co-ops (Anderson, Frenette, and Webster 1977). Already at the end of the 1960s, two lobster fishermen co-ops tried to help the mossers, but the experiment was stopped after the managers of the co-ops found it very difficult to market the seaweed stocks. In fact, although competing among themselves for raw materials, the three corporate buyers were all united in their efforts not to buy from the collective enterprise of the mossers.

In 1972, a new co-op, wholly devoted to seaweed, was founded. Beside marketing the stocks, the new initiative provided the mossers with drying grounds and storage facilities. The co-op, which had only a minimum starting capital put together by its members, was supported by modest public loans and grants supplied by the Provincial Department of Development. In 1972 and 1973, the co-op fared rather well as it managed to place sales in England, France, and Japan and redistributed the profits among its members. This result depended on the skills of the co-op manager but also on the fact that, seaweed demand being low, the corporations did not try to jeopardize the new initiative. In 1974, however, demand skyrocketed, and the co-op risked failure. Attracted by soaring prices, by a slackened attention to seaweed quality, and by giveaways in goods and services, most people left the co-op and sold to corporations. Despite this fact, by the end of the year the co-op had gathered 600,000 pounds. This stock however remained unsold for two years because the main buyers, after having been approached, refused to buy. This refusal was partly the result of market conditions, partly a conscious attempt to marginalize the co-op. Apparently, in fact, MCI maneuvered to supply directly Peyrefitte-Auby (which had just begun buying Canadian seaweed) rather than allow a deal between the French and the co-op.

In 1974, the Maritimes attained their biggest production ever. The following year, however, the market recorded a radical slump. Corporations, having filled their warehouses with the product hoarded the previous year, sharply downscaled buying activities while prices dropped. These facts had disastrous effects on local seaweed production, which fell by 28 percent and went down again by another 13 percent in 1976. By 1977 mossing was already a memory in Newfoundland, and in New Brunswick only a few harvesters continued their activities. This downgrading process persisted in the 1980s. In 1984 mossing finally disappeared in New Brunswick while in Nova Scotia quantities were soaring at the productive levels of 1953. Only Prince Edward Island continued to have some substantial production although at levels that were on average half of those of ten years before.

In the letter cited at the beginning of this chapter, the president of MCI presented his corporation's decision to buy less Canadian seaweed as a necessary adaptation to fateful circumstances created by the law of supply and demand. Appearances apart, however, this development was not only expected by the corporations, it was the outcome of years of effort on their part. To understand why, we have to go back almost two decades to follow distant developments which became clear to the Canadian mossers only after they had had to pay for their effects.

CORPORATE PLOTS

During the 1950s, as carrageenan uses began to expand at the expense of other colloids and new applications were discovered, the marketing potential of carrageenan became clear. The processors however were also concerned that if costs were not contained and a certain stability in the supply of raw materials was not obtained, such potentials would not be fully achieved (Deveau 1975). The industry was also worried, after the experience with other phycocolloids such as agar, about the potential damages produced by an overexploitation of seaweed beds. To solve these problems, the industry moved in two directions. First, it sought to diversify seaweed sources. Seas all over the world were scouted, and new types of carrageenophytes were discovered in quantity in Chile and in Southeast Asia. Second, they spurred research to improve knowledge of the ecological-reproductive requirements of seaweed.

But, the corporations realized that, unless the problem of raw materials supply was permanently solved, carrageenan market potentials could never be fully achieved. Thus the processors began testing the possibility of seaweed cultivation. Experiments began at the University of Hawaii in the early 1960s financed by MCI and by a matching grant from the U.S. Sea Grant (Laite and Ricohermoso 1981). Results were promising; to achieve commercial relevance, however, they needed to be replicated on a large scale. The point was to find around the world areas ecologically and infrastructurally fit for the commercial operation.

The ideal place was found in the southern Philippines, more exactly in the Sulu Archipelago, where MCPI, the local subsidiary of MCI, and two of the European refiners—Peyrefitte-Auby and GENU (the affiliate of KPF)—had already begun buying gathered seaweed several years earlier. The area offered a native strain of red seaweed called *Euchema* which is as strong and productive as the Irish moss and had vast reefs with all the requirements indicated by the Hawaii experiments. Last, but not least, the Samal and Tausug people living in the area had been familiar with practices of merchant capitalism for centuries and were facing problems caused by overpopulation and depletion of traditional marine resources.

THE PHILIPPINES: IMPLEMENTING
COMMERCIAL CULTIVATION

Seaweed cultivation entails a very simple technology. The activity is carried out mostly during low tide[2] and intensifies in the phase of the lunar month when the daytime low tide is longer. After the first experiments, the method that proved to be the most effective was the so-called hanging method which I observed in 1988 and 1989 during my fieldwork in Balabac, Palawan.[3] Small seaweed cuttings are tied 30 centimeters from each other along a 10- to 15-meter fishing line. Once the operation, which is performed inland, ends, the lines are brought to the planting site, optimally a clean reef with a low-tide water level of about 1 meter. There the lines are tied in rows about 50 centimeters from the sea floor between two wooden stakes. At the next low-tide phase, the seaweed is harvested, recut, tied, and replanted again. These operations repeat each lunar month and become particularly intense during the rainy season when seaweed grows faster than in other periods. Notwithstanding its dependence on sidereal and seasonal cycles, seaweed cultivation remains essentially a continuous, all-year-round activity.

Depending on environment fertility and season, seaweed growth rate may range between 2 and 10 percent a day; the doubling of the crop size in one lunar month is considered a good average result (Doty 1987). In order to increase final outputs, farmers do not sell their product after each harvest. Instead they replant the thalli as many times as the plants' resistance and their labor capacity, space availability, and subsistence needs allow. As I observed in Balabac, planters tend to wait between three and four lunar cycles before drying and selling the bulk of their plants and starting a new cultivation series. Thus, in less than four months, a planter can multiply his original seaweed pool between eight and sixteen times. This phenomenal productivity, coupled with relatively small starting capital requirements, makes seaweed cultivation a very attractive activity. A planter can initiate operations with a simple dug out. Although in this phase productivity will be far below the maximum standards, he still will be able, unless prices drop dramatically, to purchase a used pump-boat in two years. Once this implement is obtained, a planter's productivity increases enormously as he becomes able to speed up his operations both by reaching planting grounds faster and by increasing the volume of seaweed he can plant or harvest in one low-tide day.

In remote areas, seaweed trade was carried out by merchants dealing with various products (dried fish, copra, shark fins, and so on) obtained partly on the basis of *suki* (a commercial agreement by which a trader obtains produce in return for capital loans) and partly through cash exchange. The product was transported to the buying stations of the corporations or of their Filipino brokers which also bought from planters nearby. As for large traders, this intermediate marketing step was bypassed since the product went directly to the central warehouses in Cebu City or Zamboanga.

Although the biggest corporations in the Philippines keep a few plantations, seaweed cultivation is primarily a household activity. Corporate involvement in production occurred in the mid-1970s as a move to compensate for the over and undersupply crises typical of the first years of the Philippine endeavor. But, as these periodic crises became less radical, plantations soon became uncompetitive vis à vis household farms.[4] Plantations face labor and infrastructural costs that are not computed in the informal organization of a family farm, and they have more problems of labor management. This well-known principle has implications particularly dramatic for seaweed cultivation because of specific aspects of the labor process. Being determined by the lunar cyclicity of the tides, this activity entails the succession of very short periods with very high manpower demand and much longer periods in which this need is minimal. This fact created basic obstacles to a greater expansion of plantations, a kind of operation that cannot deploy a labor force flexible enough to fit the requirements of seaweed production and still maintain low operating costs. However, they did not hamper the development of a myriad of small family farms first in the Sulu Archipelago and later elsewhere in the Philippines. Not only were household labor forces elastic enough to meet the uneven demands of the new activity, the very cyclicity of this activity gave the planters time enough to pursue older productive endeavors. Although these became second to seaweed as a means to satisfy the growing living standards of the planters, they still offered subsidiary income and food particularly important in periods of low seaweed prices.

Because of its profitability, simplicity, and low labor and capital requirements, seaweed cultivation was easily taken up by Samal and Tausug settlers in Sulu. The new activity spread with phenomenal speed and intensity. The first pilot farm in Sitangkay was set by MCPI and became operative by 1971; the following year, the first shipment of cultivated seaweed left for Zamboanga City. By 1973 there were more than 100 families whose main occupation was *Euchema* cultivation; the next year, their number had risen to 1,500 (Laite and Ricohermoso 1981). Since then, the expansion of seaweed cultivation in the Philippines has been continuous, not only in Sitangkay, but also elsewhere in the Sulu Archipelago, not to mention the Visayas and Palawan. By 1978 the country had surpassed Canada as the biggest world producer of the commodity (see Table 6.1).

To be sure, this growth has not been smooth. In 1975, in fact, oversupply hit the Philippines as well as Canada. When all the corporations in the Philippines—MCPI, GENU, and Peyrefitte-Auby—stopped buying because they were already overstocked, and when, following the speculations of secondary buyers, prices at the source dropped from 2.3 to 0.3 pesos, the entire operation threatened to collapse. On that occasion hundreds of Sulu's households, that had taken up seaweed cultivation just one or two years before, abandoned the planting sites and, in the absence of solid economic alternatives, migrated to Sabah, where the oil boom was creating many low-paying jobs. Simultaneously, confronted by this disaster, all foreign corporations except MCI left the Philippines.

Unlike Canada, however, the Philippines recovered. From 1977 onward, the seaweed industry of that country not only began again to grow but also underwent basic transformations resulting in the emergence of a flourishing processing activity. To understand how this happened we have to briefly examine the events that occurred both at the level of the local procurement sector and at the level of international trade. Apart from MCPI, which initiated the Philippine farming operations and, through its own paid personnel, managed to build solid marketing links with producers, the other two multinationals relied on independent middlemen to satisfy their procurement needs (Hollembeck 1983). For newcom-

Table 6.1
Yearly Production of *Euchema Cottoni* in the Philippines and Irish Moss in Canada, 1966–1985 (in 000's dry MT)

YEAR	PHILIPPINES	CANADA
1966	0.5	5.7
1967	0.4	8.9
1968	0.1	9.7
1969	0.3	11.3
1970	0.2	11.9
1971	0.2	9.0
1972	0.3	6.1
1973	0.7	8.4
1974	10.0+	14.8
1975	1.6	8.9
1976	1.4	6.8
1977	6.0	6.3
1978	11.6	7.3
1979	16.4	6.9
1980	15.7	6.5
1981	19.5	5.5
1982	23.0	5.0
1983	19.9	4.6
1984	24.2	5.0
1985	30.9	5.5

Note: The data provided by Anderson, Frenette, and Webster (1977) are given in wet pounds; the figures of Pringle and Mathieson (1987) are in wet metric tons. I have obtained figures in this table by multiplying the figures given by these authors by 4 (the average shrinkage rate upon drying is about 75 percent). The figures in pounds were divided by 2.22 (a pound corresponds to .455 kilograms).

Sources: Philippines: Until 1978, Hollembeck (1983) and Ricohermoso-Devau (1977); from 1979, personal communication with Expedito Dublin and Veloso (1988). Canada: Until 1976, Anderson, Frenette, and Webster (1977); from 1977, Pringle and Mathieson (1987).

ers unexperienced in local marketing practices, without knowledge of local cultures, and without the necessary connections, it would have been impossible to do the job by themselves. When the hoarding frenzy of 1984 ended and the corporations stopped buying, the enormous amount of seaweed still available was purchased at slashed prices by two Philippine marketing firms which in the previous years had accumulated substantial capital operating as brokers for the Europeans. Later, these marketing firms sold their stocks in Japan at current world prices obtaining phenomenal returns.

For Japanese food and chemical processors, these deals were the break needed to enter carrageenophite markets, until then the monopoly of U.S. and European concerns. For the Philippine firms, on the other hand, the deals offered the chance of negotiating a technological transfer that would have allowed them to pass from marketing to processing. This occurred in 1977 when SHEMBERG,[5] the biggest of the Philippine firms, obtained from Japan a new, cheaper processing method and set up an experimental plant. The following year, the operation began to export low-grade carrageenan. As for the most sophisticated applications, the quality of this product, called semirefined carrageenan (SRC), could not match that of the product, called refined carrageenan (RC), manufactured in America and Europe by the big multinationals. SHEMBERG's SRC, however, was cheap and good enough to constitute a substantial challenge at the level of less demanding applications. It is thus not surprising that SHEMBERG's processing exploits prompted the reaction of MCI, the only multinational still in the Philippines, which in 1979, through MCPI, began to operate a processing facility similar in technology to that of its competitor. In 1980 the same step was taken by Marcel, the other Philippine marketing firm that played a role in the mid-1970s events. During the following decade, this step was repeated several times by other operators among them GENU, who returned to the Philippines, and a score of minor concerns. By the end of the 1980s, not only were there in the Philippines at least ten processors of SRC but also both SHEMBERG, in 1985, and Marcel, in 1989, had managed to develop RC factories.

The achievement of vertical integration of the Philippine seaweed industry has two salient features. It was reached without any substantial participation of the state, and it has produced a pool of operators very different from each other in terms of their international financial connections. Some of them were clearly affiliates of big multinationals, others were joint ventures, and at least one, SHEMBERG, seems to be entirely a Philippine operation.

POST-1975 CANADA

We have already noticed how, after the drop in demand in 1975, the Maritime seaweed industry began its inexorable decline. After 1977, as the Philippines recovered from the same event, the corporations could count on another source of raw materials—one cheaper and more stable than the Canadian. This situation gave the corporations a basic advantage vis à vis the rising demands for higher prices of the mossers who went repeatedly on strike obtaining only minimal results. As the multinationals' capacity of price control grew, the mossers operating in the most remote areas stopped their activities while others managed to survive thanks to unemployment subsidies. Another effect of corporate buying policies was the final disappearance of all the remaining intermediate buyers except the co-op, which continued to operate thanks to government help. These results were reinforced by corporate mergers that greatly decreased the competition between multinationals. In 1977 MCI was bought by FMC, a big player in the American food industry which also bought Litex, in 1978, and Stauffer, in 1979. This last acquisition, which had been previously pursued by MCI, but had been stopped by U.S. antitrust laws, became possible because another big

U.S.–based chemical conglomerate, Hercules, had bought KPF in the previous years.

Thus, if, on the one hand, corporate strategies brought about a gradual decline of mossing activities in the Maritime Provinces, on the other, they reinforced the corporations' dominance of the local raw materials markets.[6] An important aspect of this outcome is that it was partly helped by the structure of Canadian development programs. Apparently, as voices speaking in favor of the Maritime local economy (see Anderson, Frenette, and Webster 1977) lamented, the multinationals had fewer difficulties in obtaining federal development money than small local entrepreneurs. According to these sources, this happened for two main reasons. First, because Ottawa officials were too far removed from the mossers' problems, government aid privileged a cost/benefit analysis in which purely financial and commercial considerations were more important than socioeconomic ones. Second, bureaucratic complications had the effect of making the application process too slow and complicated for small investors.

As a result, the government paid little attention to various projects aiming at building extractive facilities in loco. Although highly recommended by local interests in the Maritime, such as the co-op, which considered such projects as the only way to add value to seaweed and thus raise mossers' returns, the construction of a refinery was deemed too risky since the world markets were already dominated by foreign corporations. However, in the late 1970s, both MCI and GENU, as firms incorporated in Canada, got from federal funding institutions substantial capital to undertake experimental moss cultivation. These projects were safer from the perspective of potential profits, but they were against the mossers' interests. In fact, in the Maritimes, moss cultivation entailed a technology and a socioeconomic impact entirely different from those in the Philippines where seaweed is planted in the sea with a very simple, cheap technology and can be carried on all year round as a household activity. Because Canadian climate makes mariculture unthinkable, research moved in the direction of pond cultivation. Not only is this planting method labor intensive, it is capital intensive. Its implementation needs big infrastructures and is possible only as a corporate endeavor. Thus, when Canadian seaweed cultivation finally got under way in the late 1980s, its effect was to weaken even more the already ailing mossing industry in the Maritime Provinces.

PHILIPPINES AND CANADA COMPARED

The Philippines' success over Canada has depended on several factors. The two most important are the achievement of a reliable supply of seaweed, which was obtained through the implementation of cultivation, and the differential in price paid to the direct producers in the two countries. In 1972 the price paid to farmers in Tawi-Tawi was 1.20 pesos/dry kilogram (Hollembeck 1983), or $0.16; the price paid in the Maritime Provinces was $0.03 wet pound (Anderson, Frenette, and Webster 1977), roughly 26.6 cents/dry kilogram.[7] Considering the standard cost of living and the available alternatives to seaweed cultivation in the Philippine seaweed-producing areas, that price was considered more than sufficient to attract hundreds of households into the new activity.

Both Canada and the Philippines suffered great damages during the 1975 crisis of oversupply, but, in 1977, when the crisis finally abated, the outcomes for Filipinos and Canadians were very different. In the case of the Philippines, not only did production resume expanding in several new islands and provinces, but also new important developments favoring local operators against the foreigners occurred. On the one hand local marketing firms that in the past had worked as middlemen for the American and European corporations became exporters on their own and simultaneously obtained the leadership of the local procurement sector. On the other, the Philippines managed to develop its own refining facilities. This outcome depended on a combination of factors specific to the Philippine situation. These factors can be synthesized as follows: first, the local product had a low price by world standards; second, the multinationals had weak control of the raw material procurement sector; and third, Philippine entrepreneurs had access to alternative sources of capital.

What happened in Canada seems to be the negative mirror image of what occurred in the Philippines. In fact, in Canada, the recovery of the world markets after the 1974 oversupply crisis did not bring any stable improvement in local seaweed production. Apart from a mild recovery in 1978, in the following years, the amount of Canadian seaweed fell constantly. Only the corporations and the foreign interests represented by them obtained relative benefits from this situation as they managed to obtain an even bigger control of the local product than before. The crisis in fact wiped out from the scene all the independent buyers except the co-op.

In the Canadian case, the factors that played a role were nearly the opposite of those found in the Philippine situation. They can be summarized as follows: First, high prices limited the marketing potential of Canadian seaweed; second, the multinationals had obtained a nearly complete control of the procurement sector; and third, although capital and technology were more than welcomed by local operators, neither foreigners nor the Canadian government decided to invest directly in processing facilities. In addition, unlike in the Philippines, where the state did not participate in the early development of the seaweed industry, the Canadian state gave some degree of support to the multinationals.

CONCLUSIONS

The opposite fates of the Canadian and the Philippine seaweed industries must be interpreted in terms of a progressive interplay between changing global and local factors and in terms of the influence on that interplay of corporate strategies. As colloids like carrageenan became an essential if "invisible" component of the durable foods at the top of the new agro-food chain, their demand increased enormously. In the case of colloids and thickeners, however, demand increase can hardly be followed by corresponding price increases. As was noted at the beginning of this chapter, such substances can easily substitute for each other with the result that their commercial viability depends on a subtle balance between technical properties and costs. The outcome of these marketing conditions has been that manufacturers, in order to keep prices low and to increase their product versatility, pursued with particular intensity global sourcing and research. This process was facilitated by the waning role of state apparatuses in protecting national interests.

Corporate practices, however, had different results and successes in different places. In this sense, the variable combinations of local factors, such as ecology, technology available, mode of production, degree of corporate control of raw materials, and competition between operators, have played a very important role. In Canada, the combined outcome of local conditions and global sourcing had the effect of simultaneously tightening corporate control of raw materials and decreasing local production. In the Philippines, the opposite happened as local conditions suited the successful establishment of seaweed cultivation but limited in the long run corporate access to productive outputs.

This last development emphasizes both the role of corporate strategizing and its uncertainties. Notwithstanding advantages in terms of resources, technological know-how, positioning in the market, and knowledge of global trends, multinationals still act in terms of a "bounded rationality" (Evans 1979) where market conditions are often beyond their control. Important in this sense are the unintentional outcomes of corporate competition. For instance, although we can assume that multinationals are interested in maintaining the operators in the periphery in a situation of dependency, by competing with each other, they may open spaces in which these periphery operators may maneuver a better position for themselves. This surely happened in the Philippines. MCI's establishment of seaweed cultivation aimed at expanding and stabilizing the corporation's raw materials requirements and simultaneously achieving lasting advantages both vis à vis direct producers and other competing corporations. However, because of unexpected local results, most importantly the phenomenal response to the MCI initiative of the population of the Sulu Archipelago, what the corporation wanted occurred too rapidly. As seaweed cultivation spread and the produce reached levels far beyond the absorption capacity of MCI, not only was the corporation unable to monopolize the results of its efforts—both its Danish and French competitors managed to fill all their needs of carrageenophytes for at least the next two years—but also a world overproduction crisis occurred. The occasion favored the emergence of Philippine exporters and, later, thanks to the involvement of Japanese capital and technology, of Philippine processors.

A major trait of these outcomes was the specific timing that coordinated local productive and commercial developments, wider trends in world agro-food markets, and other aspects of the global economy—particularly the emergence, in the mid-1970s, of Japan as a major world economic player. As Japan became a potential buyer of carrageenophytes, Canadian seaweed was already too scarce and too tightly controlled by the multinationals operating in the area to justify a strong Japanese interest in local produce. Philippine seaweed, instead, was cheap, plentiful, and available. This allowed Japan's entrance in the carrageen-ophytes markets, and it set the basis for a transfer of technology and capital that permitted the establishment of Philippine processing.

NOTES

1. Agar, a product better known to the general public than carrageenan, was the first phycocolloid to be synthesized in pure form and the first to become an industrial com-modity. Derived from different kinds of red seaweed than carrageenan, mostly *Gigartina* and *Gelidium*, agar has been used in a vast array of applications ranging from food additives to a cultural medium of microbiological research. Because the supply of its raw materials has not yet been stabilized by cultivation and because of its high extinction costs, agar is particularly expensive. For this reason, in the last forty years, its less demanding applications have been slowly taken over by other cheaper phycocolloids. This, however, has happened only in a small degree for utilizations demanding maximum gel strength and purity.

2. If the water level forces the planters to put their heads under water, planting and harvesting become very difficult, if not impossible, tasks.

3. Fieldwork in the Philippines was carried out between December 1987 and June 1989 and was financed by a grant of the Social Science Research Council, Southeast Asia Committee and the H. Luce Foundation.

4. According to Lim and Porse (1981), at that time the production cost of *Euchema* was $0.34 per kilogram in a corporate plantation and only $0.20 in a household-managed farm. Today the major processors continue to maintain corporate plantations for three reasons: they function as research stations; they help to promote a solid corporate image attractive to prospective customers; and, probably most important, they can be devoted to the cultivation of particular strains of seaweed, such as *Euchema spinosum*. This kind of seaweed is essential for the preparation of specific types of carrageenan and usually has a high market value, but it is not favored by the farmers because of its slow growth rate.

5. Notwithstanding its German sounding name, the corporation is a 100 percent Phil-ippine operation, owned by the Dakay family, one of the biggest industrial and com-mercial groups of Cebu City. Its name is an acronym formed by the initials of the names of the children of the old Mr. Dakay.

6. Cost and irregular supply notwithstanding, the multinational corporations main-tained an interest in Irish moss because the plant is an essential component of one type of carrageenan that cannot be refined from other kinds of seaweed.

7. See note in Table 6.1 for the conversion parameters used. Anderson, Frenette, and Webster (1977) report 1972 FOB prices from the Philippines and from Canada, which were, respectively, $270 to $300 and $500 to $550 per dried metric ton.

REFERENCES

Anderson, N., E. Frenette, and G. Webster. 1977. *Global Village, Global Pillage: Irish Moss from P.E.I. in the World Market.* Charlottetown, P.E.I.: Report on Social Action Committee for Roman Catholic Diocese of Charlottetown.

Cardoso, F. H. 1977. "The Consumption of Dependency Theory in the United States." *Latin American Research Review* 12: 7–24.

Deveau, L. 1975. "The Seaweed Industry in the Maritime Provinces." In *Proceedings of the Bras d'Or Lakes Acquaculture Conference.* College of Cape Breton Press.

Doty, M. S. 1987. "The Production and Use of Euchema." In *Case Studies of Seven Commercial Seaweed Resources,* edited by M. S. Doty, J. M. Cady, and B. Santelices. Technical Paper 281. Rome: FAO Fish.

Evans, P. 1979. *Dependent Development: The Alliance of Multinational, State and Local Capital in Brazil.* Princeton, N.J.: Princeton University Press.

Hollembeck, L. F. 1983. "An Economic Assessment of the Philippine Euchema Seaweed Industry with Implications for Future Policy." Master's thesis, University of Hawaii.

Laite, P., and M. Ricohermoso. 1981. "Revolutionary Impact of *Euchema* Cultivation in the South China Sea on the Carrageenan Industry." In *Proceedings of the 10th International Seaweed Symposium,* edited by Tore Levring. Berlin: Walter de Gruyter.

Lim, J. R., and H. Porse. 1981. "Breakthrough in the Commercial Culture of *Euchema Spinosum* in Northern Bohol, Philippines." In *Proceedings of the 10th International Seaweed Symposium,* edited by Tore Levring. Berlin: Walter de Gruyter.

Mintz, S. W. 1977. "The So-Called World System: Local Initiative and Local Response." *Dialectical Anthropology* 4: 253–70.

Pringle, J. 1975. "The Marine Plant Industry—Commercially Important Species and Resource Management." In *Proceedings of the Bras d'Or Lakes Aquaculture Conference.* College of Cape Breton Press.

Pringle, J., and A. C. Mathieson. 1987. "Chondrus Crispus Stackhouse." In *Case Studies of Seven Commercial Seaweed Resources,* edited by M. S. Doty, J. F. Caddy, and B. Santelices. Rome: FAO.

Stoloff, L. 1949. "Irish Moss—From an Art to an Industry." *Economic Botany* 3.

Veloso, A. R. 1988. *Euchema Farming in the Philippines.* Paper presented at the Workshop on Mariculture, Silliman University, Dumaguete City, October 15–22.

7

Vines and Wines in the World-Economy

Roberto P. Korzeniewicz, Walter Goldfrank, and Miguel E. Korzeniewicz

This chapter focuses on Argentina and Chile to elucidate the nature of recent changes in the fresh fruit and vegetable global commodity chain (FFVGCC) and the wine global commodity chain (WGCC). This exercise will allow us to clarify the characteristics of successful export strategies in Argentina and Chile and to reevaluate current changes in the study of development. The first section provides an overview of the development of fresh fruit and wine exports from Argentina and Chile, as well as a synthesis of the main variables—state policies, labor, and entrepreneurial innovation—generally used to explain the success of export strategies in Chile and their failure in Argentina. The second section of the chapter explores the contribution of world-system analysis to a better understanding of the processes at hand.

EXPORT GROWTH IN ARGENTINA AND CHILE DURING THE 1980s AND 1990s

To avoid the constraints and stagnation that were perceived to be associated with import-substitution industrialization (ISI), policy makers in Latin America have pursued a new model of development over the past decade. The general features of this model have by now become quite apparent. State policies have aimed to dismantle most forms of corporate and state regulation of the market to allow unfettered market forces to become an engine of economic growth. In pursuing this objective, policy makers in Latin America have turned to Chile as a potential model of economic growth.

Chile's high rates of economic growth have been closely tied to the successful implementation of an export strategy. Enterprises in both Argentina and Chile have increasingly sought to become important players in the FFVGCC and the WGCC, but the degree to which they have achieved this objective (as measured, for example, by the level of exports attained by these enterprises) shows considerable differences between the two nations. In the area of fresh fruit exports, the difference between Argentina and Chile has been simply phenomenal, as suggested by Table 7.1.[1]

Starting in the late 1970s, enterprises in Chile greatly expanded fresh fruit exports especially after the agricultural crisis of 1983. Although enterprises in Chile had a longer trajectory exporting grapes than their counterparts in Argentina, this was not the case with other fresh fruits (such as apples) and wine. In the case of wine, enterprises in both Chile and Argentina faced declining international consumption during the 1980s, but those in Chile were more successful in responding to these constraints by greatly expanding exports. By the 1990s, each of these commodities shows significantly greater exports in Chile than in Argentina.

Furthermore, enterprises in Argentina and Chile have aimed at different export markets. The composition of exports of wine and must from Argentina suggests that enterprises in this country sought to capture niches for lower priced wines (such as wine in bulk). In Chile, on the other hand, enterprises have aimed at higher value niches in world markets, concentrating on exports of bottled (rather than bulk) wine. Of course, higher value is in relative terms, for the sales of virtually all semiperipheral wines in core nations tend to be limited to the lower end of the premium wine category, where semiperipheral wines can aim to deliver relatively good quality at low prices and without the prestige associated with core wine producers (e.g., California, France, Italy, or even Spain).

Observers have tended to emphasize three variables in attempting to explain the comparative success of Chile and the failure of export strategies in Argentina: appropriate and stable state policies, the existence of a disciplined and inexpensive labor force, and an innovating entrepreneurial culture.

State Policies

The first variable that is frequently used to explain successful export strategies involves patterns of state action and regulation. In the case of Chile, multiple authors (Cruz and Leiva 1982; Sáez 1986; Falabella 1988; Gomez and Echenique 1991; Rivera 1991; Gomez 1991) have suggested that state development strategies during the 1960s and early 1970s were significant in creating a foundation of experience and resources that could later provide a basis for the rapid growth of fresh fruit and vegetable (FFV) exports.[2] Furthermore, the transfer of state-owned land after the coup of 1973 favored the fruit sector (Cruz and Leiva 1982, 14). In addition, "Chile has developed an extremely aggressive commercial policy in terms of macroeconomic measures such as high exchange rates and economic openness" (Juri 1991, 33).

Table 7.1
Exports of Fresh Fruit, Apples, Grapes, and Wine from Argentina and Chile, 1970–1991 (thousands of dollars)

	Fresh Fruit Argentina	Fresh Fruit Chile	Apples Argentina	Apples Chile	Grapes Argentina	Grapes Chile	Wine Argentina	Wine Chile
1970			89,151	15,362	n/a	n/a	477	1,840
1971			93,616	18,205	n/a	n/a	514	1,806
1972			117,797	19,674	n/a	n/a	730	2,132
1973			151,661	42,168	n/a	n/a	2,645	2,901
1974			74,275	5,517	n/a	n/a	4,124	3,599
1975			89,151	15,362	n/a	n/a	3,286	3,920
1976			93,616	19,500	971	15,249	6,558	7,843
1977			117,797		1,926	19,554	10,105	7,589
1978					1,965	29,649	16,146	11,895
1979			143,833	44,524	1,168	46,086	9,301	22,140
1980			193,203	65,969	5,641	41,047	6,053	20,235
1981			148,478	76,100	3,245	69,756	5,957	15,690
1982			159,620	142,718	3,473	141,782	6,550	12,521
1983			128,958	126,230	4,023	232,272	6,995	19,003
1984			55,212	148,307	715	338,126	7,347	20,411
1985			57,494	148,853	490	431,333	5,894	21,979
1986			65,490	253,858	929	509,067	6,571	26,535
1987			66,850	216,769	1,942	335,900	8,819	17,183
1988			56,466	230,013	3,486	389,068	10,869	17,045
1989			54,714	348,669	5,655	597,917	14,707	28,240
1990			74,778	364,854	7,607	639,228	15,235	51,568
1991							17497	84144

Sources: Elaborated on the basis of data from United Nations, Department of Economic and Social Affairs, International Trade Statistics Yearbook, 1974-90; for 1990 and 1991 data on wine from Chile, Asociación de Exportadores y Embotelladores de Vinos, 1992; for 1990 and 1991 data on wine from Argentina, Instituto Nacional de Vitivinicultura, 1992.

In the case of Argentina, state policies are blamed for having constrained market forces, hampering the growth of exports. For example, Juri (1991, 9) has attributed the stagnation of Argentine exports to "the lack of an institutional framework" that resulted in turn from "inadequate macro and microeconomic policies that imprinted the economy with a level of uncertainty, instability and inefficiency that is incompatible with . . . production for exports." Among these inadequate areas, Juri includes inflation, interest rates, taxes, exchange rates, excessive state regulation, and inefficient state services (e.g., transportation and communications). Fluctuating exchange rates have had a particularly detrimental effect on exports. In the case of wine production, for example, high exchange rates have often led domestic prices to be higher than international prices.[3] In short, according to this line of interpretation, "Argentina has lacked the capacity to deliver fruits and vegetables in competitive conditions from the fields to the market, not so much from managerial incompetence at the entrepreneurial level as from the Argentina institutional framework" (Juri 1991, 49).

We should note that the broad consensus over the need to reduce state regulation of market forces tends to break down over the issue of trade regulation. Recent efforts to develop a trade agreement between Argentina and Chile came to a halt over the issue of the possible transit of Argentine fruit through Chile. Claiming that the transit of Argentine fresh fruit could introduce pests into their country, export enterprises and fresh fruit growers in Chile blocked the relevant sections of the trade agreement.[4]

Labor

The characteristics of the labor force are frequently cited as a crucial competitive advantage, particularly in the seemingly labor-intensive production of FFV.[5] Over the last two decades, social relations in agriculture in both Argentina and Chile have undergone significant transformations, involving the displacement of tenant farmers and permanent wage workers by seasonal workers, the incorporation of women and youth as a significant component of this seasonal labor force, and the growing importance of contractual arrangements between independent producers and marketing firms. But whereas these transformations are perceived to have facilitated the expansion of fresh fruit production in Chile, the characteristics of the labor force (particularly in regard to costs) continue to be perceived as an obstacle to the growth of fresh fruit exports in Argentina.

In Chile, agrarian reforms undertaken between 1964 and 1973 undid traditional relations between landowners and tenants, producing both a larger number of agricultural wage workers and new "modern entrepreneurs" (Falabella 1988, 2). These modern entrepreneurs gained strength after the 1973 military coup. The period from 1975 to 1983 was characterized by strong competitive pressures that squeezed indebted farms and led to land concentration (Gomez and Echenique 1991, 46). During this period, the peasantry and rural labor force became further impoverished (Falabella 1988, 4).[6] Thereafter, government incentives and available capital resources helped promote the formation of a new power bloc in agriculture, dominated by "medium enterprises with a high capital density" (Sáez 1986, 13, 26; see also Cruz and Leiva 1982, 59). This shift transformed capital-labor relations: "[I]nstead of a secure relationship, based on the loyalty of the tenant, the patron now established, as a product of his mistrust, punctual and seasonal relations—uncertain and insecure for the worker—with most of the labor force" (Falabella 1988, 4). Thus, "agricultural enterprises have reduced their obligation to ensure the subsistence and reproduction of the necessary labor force, lowering their fixed costs and incorporating into their commercial domain the land resources freed by the resident peasants. The old pyramid of command hierarchies has collapsed" (Gomez and Echenique 1991, 65).

The agricultural labor force in Chile in 1986–1987 was estimated to involve 796,000 people, or 18.5 percent of the total paid labor force (Gomez and Echenique 1991, 55). The number of seasonal agricultural wage workers increased from 147,000 in 1964–1965, to 198,000 by 1975–1976, and 300,000 by 1986–1987; during these same years, on the other hand, the number of permanent wage workers declined from 208,000 to 161,000, and to 120,000 by 1986–1987 (Gomez and Echenique 1991, 64; Cruz and Leiva 1982, 86, 108).

Many of these seasonal workers (the packing season peaks between December and March) are urban and rural women, for whom this employment may represent their first job at a young age or an alternative to domestic (waged or unwaged) work (Medel R., Olivos M., and Riquelme G. 1989).[7] One set of authors suggests that, in one particular area of fresh fruit production, women account for 28.3 percent of the labor force in field work and for 76.1 percent of the packing workers (Medel R., Olivos M., and Riquelme G. 1989, 52). Falabella (1988, 1) has indicated that, in the Valle de Aconcagua in 1988, women accounted for over 50 percent of the labor force. There data available are limited on the gender composition of the labor force or even on such issues as wages or labor unrest, hampering a more systematic evaluation or comparison.

Much of the literature deals with this incorporation of women into the fresh fruit labor force in terms of the exploitative conditions prevailing in the area. But participants often suggest a more complex situation. In the study carried out by Medel R., Olivos M., and Riquelme G. (1989, 35), for example, a female homemaker (forty-four years old) indicated that wage labor "is also like a therapy. Because there one really forgets that she has a house, that she has children." Female informants often emphasized that work allowed them to enhance their social life, indicating, for example, that during the off-season they missed "the affective and intimate atmosphere generated in the work space" (Medel R., Olivos M., and Riquelme G. 1989, 36). Another female worker, a twenty-six-year-old single mother, indicated to the same authors that "going to [work] I felt as 'yo mujer,' not 'yo madre,' but 'yo mujer.' . . . That is to say, to get the pleasure of being with my friends, being with my amiga. . . . When I board the bus (returning home) I am thinking how will I look again as madre" (Medel R., Olivos M., and Riquelme G. 1989, 37).[8] Falabella (1988, 7) has indicated that gender relations in fruit production are characterized by "a level of gender equality (women earn even more money and due to this have greater status than men) never seen in the countryside.'' The incorporation of women into the wage labor force hence leads to changes in the internal organization of households and their own relationship to enterprises and the state (for example, by intensifying anxiety and concern around new issues such as child care).

Data on wages and working conditions among these workers are also scarce. Some authors have emphasized the poor quality of these arrangements for workers. For Gomez (1991, 17), the Chilean success in exports was based not only on the skills of entrepreneurs, but also on ''the sacrifice of thousands of seasonal wage workers in the fruit and forestry [sectors].'' Similarly, earlier in the 1980s, Cruz and Leiva (1982, 16) suggested that ''the high rate of unemployment and the decrease of the level of wages that prevailed until 1973'' provided a strong comparative advantage of enterprises in Chile.

Nonetheless, as opposed to other forms of seasonal agricultural employment in Chile (e.g., beet production, wood harvesting), fruit production requires greater specialization and care, leading to the development of more stable labor arrangements (Falabella 1988, 5). Sáez (1986, 16) indicates that, in some areas of Chile, the rapid expansion of fresh fruit production led to labor shortages and higher wages, and as economic growth picked up generally in the later 1980s, wages (if not yet working conditions) seem to have improved considerably. For example, rates of unemployment in agriculture as a whole rose above 15 percent around 1983 but subsequently declined, to go below 8 percent by the late 1980s (Gomez and Echenique 1991, 57). In 1988, Gomez and Echenique (1991, 77) noted that ''during the last season of harvests in the central zone, agricultural wages doubled and occasionally tripled the minimum wage, achieving in certain areas levels close to 800 pesos (c. US$ 3.00) a day, a level higher than the wages of unskilled workers in construction and some industrial sectors.'' However, considerable regional variation exists, for ''the price of the labor force (as any other commodity) is determined by supply and demand in the regional labor markets'' (Gomez and Echenique 1991, 83).[9]

Some of the features of this broad transformation of agricultural production relations over recent decades have also been observed in Argentina. Contract agriculture in fresh fruit production was promoted by large enterprises during the 1960s to enhance their control of the labor process while minimizing their risks of production (Mendez Boaglia and Bonifacio 1992, 2). As in Chile, these changes were accompanied by growing wage differentiation, replacement of workers by machinery, and the appearance of seasonal workers on a mass basis (1992, 3). Moreover, rural workers in Argentina echo their counterparts in Chile by suggesting that, over the past two decades, ''the most substantial change is the disappearance of the figure of the patron and his replacement by tecnicos'' (1992, 12). As in Chile, the displacement of the traditional patron by a new managerial stratum became particularly pronounced after the 1980s, and it has been most clearly felt in the larger units of production. But despite these changes, the characteristics of the labor force continue to be perceived as a source of strong competitive disadvantage by enterprises exporting fresh fruit in

Argentina. In particular, business representatives continue to target labor short-ages and high wages as major constraints upon exports.[10]

Entrepreneurial Innovation

A third variable has involved the ability of business firms to adopt necessary innovations in production and marketing. The adoption of these innovations has shaped the ability of firms in Argentina and Chile to compete effectively in the international (and domestic) markets for wine and fresh fruit.

High technology and investment have come to characterize each step of the production and marketing of wine and agricultural specialty commodities. As indicated by Collins (1991, 4), "[T]he production of fruits and vegetables is a relatively high-risk enterprise," and technological changes are directed at low-ering these risks. Technological innovation includes the type of seedlings and soil preparation used in production, bioengineering, pest controls, and handling and packaging.[11] These technologies are expensive, and their cost restricts access to these innovations.[12] The incorporation of contractual arrangements between producers and marketing firms has tended to enhance the bargaining power of large firms vis-à-vis independent producers in both Argentina and Chile. In Chile, these arrangements involve a price for the producers agreed between the two sides, and a charge by the marketing firm of from 8 to 12 percent of the final sale; on the basis of this agreement, the marketing firm advances credit and technical assistance in production, sets technical norms for production, and carries out quality controls at each step in the process (Sáez 1986, 17–18). In this manner, marketing firms have enhanced their bargaining power vis-à-vis independent growers.

In Argentina, efforts to mechanize in the 1960s led small and medium fresh fruit producers into debt, although high international prices for their commodi-ties allowed these producers to subsist until 1974 (Castello, Pacenza de Del Campo and Izurieta 1990: 217). After 1974, however, falling prices and the rapid growth of the fresh fruit supply undermined small and medium producers, providing opportunities for large "integrated producers" who bought land, had easier access to (and could finance) new technologies, and enhanced their bar-gaining power as marketers (Castello, Pacenza de Del Franco, and Izurieta 1990, 218).[13] Thus, "most of the large fresh fruit enterprises became integrated back-wards . . . beginning with the export business to end up gaining access to the production of the raw material" (Castello, Pacenza de Del Campo, and Izurieta 1990, 222). As in Chile, these "integrated" firms gained considerable bargain-ing power in relation to independent growers.

We should note here that these changes do not mean that large enterprises have totally displaced small and medium firms. In both Argentina and Chile, there is considerable persistence of partially integrated producers (cooperatives), independent growers, small farms, and so forth. Along similar lines, although focusing on labor relations in agriculture, Collins (1991, 2) has indicated that the production relations through which global agriculture is organized can neither be deduced from theories of agrarian transition nor from the characteristics of crops and technologies. Production relations are generated out of historically specific social processes in which agribusiness firms seek to acquire and discipline labor in accordance with their needs.

In other words, a multiplicity of sometimes conflicting objectives, opportunities, and constraints affects the institutional strategy of individual firms, creating great heterogeneity within agricultural commodity chains as a whole.[14]

By increasing productivity, the adoption of technological innovations also tended to enhance the bargaining power of business firms vis-à-vis labor. In Argentina, for example, the adoption of new planting methods (*espaldera*) in the 1960s allowed harvests to be conducted with 23 percent less labor and with 150 percent greater productivity than the traditional methods had provided (Mendez Boaglia and Bonifacio 1992, 7–8).

Finally, the aim of these technological changes has not only been to achieve greater productivity and to reduce labor costs, they have often enhanced the bargaining power of firms vis-à-vis their consumers. For example, the installation of cold capacity allowed firms in Rio Negro, Argentina, to better withstand market fluctuations and low prices, enhancing the bargaining power of regional producers vis-à-vis the cold packing houses of Buenos Aires (Mendez Boaglia and Bonifacio 1992, 7–8).

Overall, technical innovations were more broadly adopted in Chile (in both fresh fruit and wine production) than in Argentina. In the case of fresh fruit production in Argentina, the adoption of these innovations has been practically limited to the Rio Negro region—which accounted for 73.2 percent of the apple and 67.7 percent of the pear harvests in the 1980s (Castello, Pacenza de Del Franco, and Izurieta 1990, 215).[15] The cold systems existing in areas such as Mendoza are inadequate, and the same problem characterizes most ports, for only San Antonio Este (servicing Rio Negro) has appropriate cold systems (1990, 243). Even in the Rio Negro, "controlled atmosphere" systems accounted for only 3 percent of the cold capacity in the early 1980s (Manzanal and Rofman 1989, 131). Similarly, the adoption of the latest technologies in wine production in Argentina has been limited in scope and restricted to a few firms.

This has been the case not only of technological innovations in production, but also of organizational innovations in marketing and distribution. In comparison to their counterparts in Chile, Argentine export enterprises have lacked effective commercial and marketing networks, as well as basic knowledge of the relevant markets. They have experienced problems securing appropriate inputs for export sales and problems developing the type of quality required by export markets.

Although the limited adoption of appropriate innovations in Argentina has sometimes been attributed to unstable state policies, the ability to rely on a larger internal market has played an important role as well. For example, when fresh fruit exporters faced a recession in 1979–1981 (due to falling demand and prices), they were able to shift their product to the internal market (Castello, Pacenza de Del Franco, and Izurieta 1990, 219). Thus, in fresh fruit production for the 1980s as a whole, the internal market accounted for 43.8 percent of apples and 39.1 percent of pears, industrial processing accounted for 33.1 percent of apples and 23.9 percent of pears, and exports accounted merely for 23.1 percent of apples and 37 percent of pears (1990, 224). Likewise, wine producers in Argentina "had an ensured deal in the domestic market, and within the reigning commercial framework they achieved satisfactory profits" (Manzanal and Rofman 1989, 207). This ability to rely on the domestic market has certainly been a considerable variable accounting for the different export strategies adopted by enterprises in Argentina and Chile.

Others have attributed differences between the two countries to entrepreneurial culture. For example, Sáez (1986, 26) argues that, in Chile, "we find an entrepreneurial identity that delineates a new cultural ethos." Likewise, although focusing on a different case, Morawetz (1981: 33) has suggested that "because most measures to increase productivity . . . involve changing ways of doing and organizing things rather than introducing new machinery, the efficiency of management, especially at the middle level, tends to be crucial" (Morawetz 1981, 86).

Although each of the variables reviewed in the preceding pages has played a role in shaping the contrasting trajectories of export strategies in Argentina and Chile, their character and interrelationship can be more fully developed by situating them within a world-systems framework. This is the purpose of the next section.

THE SPATIAL LOCATION OF VINES AND WINES IN THE WORLD-ECONOMY

Within more traditional approaches to economic development, commodities are still often classified in terms of the degree of processing or manufacturing involved in their production. Generally, less processed commodities are assumed to command lower levels of wealth and hence to generate less development. Until recently, these traditional approaches expected core nations to specialize in the production of more processed commodities, and peripheral nations to be concentrated in the production of raw materials and less processed commodities. World trade, finally, was expected to reflect these types of specialization, both in regard to the type of commodities exported/imported by each set of nations, and in regard to net flows of wealth among these nations.

We critically evaluate these premises by using world data to identify the spatial boundaries in the production, export, and import of table grapes and wines, commodities that sometimes continue to be depicted as characteristic forms of the low-wage, low value-added activities that characterize peripheral and semiperipheral production. But, in fact, we are dealing here with agricultural specialty commodities that can be "considered high-value because of their higher production and marketing costs and limited production regions and storability which lead to relatively higher prices per unit than many bulk agricultural commodities such as wheat" (Buckley 1990, 1). By taking these characteristics into consideration, a new perspective can be cast on the contrasting trajectories of export strategies in Argentina and Chile.

As a preliminary step, to illustrate these characteristics, Table 7.2 presents data on exports for two sets of commodities: fresh fruit and vegetables, and wine.[16] The data reveal that, in the export of both sets of commodities, there is a significant presence of core nations. In the case of FFV, China, Morocco, and Honduras appear to be important exporting nations, but exports are far greater for the Netherlands, Italy, the United States, and France (with Spain, these latter countries accounted for over half of FFVGCC world exports in the late 1980s).[17] Similarly, in the case of wine, the exports of France and Italy alone account for well over half of all world exports and are far more significant than the production of peripheral nations. The proportion of exports accounted for by core nations would change if trade within the European Economic Community were to be excluded from the data, but both the FFVGCC and the WGCC would continue to be characterized by a prevalence of core exports.

The data in Table 7.3 suggest that much of the consumption through imports of these two sets of commodities expectedly involves relatively wealthier nations.[18] In fact, from the point of view of demand, much of the growth in the FFVGCC and the WGCC was tied to the growing affluence and—particularly for FFV, but also for wine—the health concerns of consumers in core areas of the world-economy (Friedland 1991, 8; Goldfrank 1993). Some would go as far as to argue that increased consumer demand in core nations was the primary stimulus "for technological innovations that reduce two major trade limitations of these commodities: high perishability and widely varying quality and appearance" (Buckley 1990, 1). Others are more cautious about the direction of causality among these variables. Friedland (1991, 6) suggests that the seasonal extension of production and varietal expansion in FFVGCC has been driven by the combined effect of "(1) the healthy food movement and concerns about food safety, (2) the development of production capacity through the export of capital and technological expertise, and (3) the establishment of capital-intensive cool chains which maintain chilled temperatures from origin to consumption" (see also How 1991, 24).

In order to pinpoint the spatial location of exports and imports, we follow some insights from world-systems theory to depict graphically the trade relationships involved in these sets of commodities. First, we follow the procedures outlined by Arrighi and Drangel (1986) to analyze the membership of world-economic zones. Although the use of GNP per capita as an indicator of zonal positions continues to generate controversy both within and outside world-systems theory, we maintain that it provides a useful relational measure of the trajectory and composition of world economic zones. In particular, we use the

Table 7.2
Exports of Fresh Fruit and Wine and Country Share of World Exports, 1987–1988

	FRESH FRUITS AND VEGETABLES					
	Value (in $ Thousands).	World Share	Cumulative Share	Volume (in Metric Tons)	World Share	Cumulative Share
Spain	2501953.0	14.4	14.4	4489887	11.0	11.0
Netherlands	2297102.5	13.2	27.6	4395167	10.7	21.7
Italy	1669312.0	9.6	37.1	2540693	6.2	27.9
United States	1388004.5	8.0	45.1	2361770	5.8	33.6
France	1184629.5	6.8	51.9	2184408	5.3	39.0
Mexico	566149.5	3.3	55.2	1680244	4.1	43.1
Chile	546985.0	3.1	58.3	835141	2.0	45.1
South Africa	376160.0	2.2	60.5	802750	2.0	47.1
Honduras	342692.0	2.0	62.4	968478	2.4	49.4
Greece	283278.0	1.6	64.0	609357	1.5	50.9
Canada	163051.0	0.9	65.0	1483555	3.6	54.5
Costa Rica	268231.5	1.5	66.5	1075792	2.6	57.2
Israel	307312.0	1.8	68.3	627082	1.5	58.7
China	262447.0	1.5	69.8	1044828	2.5	61.2
Morocco	262587.0	1.5	71.3	690662	1.7	62.9
Other	4998389.0	28.7	100.0	15201279	37.1	100.0
Total World	17418283.5	100.0	100.0	40091090	100.0	100.0

Table 7.2 Continued

	WINE					
	Value (in $ Thousands).	World Share	Cumulative Share	Volume (in Metric Tons).	World Share	Cumulative Share
France	3322364.5	51.2	51.2	1312984	30.4	30.4
Italy	974210	15.0	66.3	1181393.5	27.4	57.8
Spain	499797.5	7.7	74.0	437078	10.1	67.9
Bulgaria	207500	3.2	77.2	197510	4.6	72.5
Portugal	325683.5	5.0	82.2	157054.5	3.6	76.1
United States	72529	1.1	83.3	52393.5	1.2	77.3
Algeria	19586	0.3	83.6	40720	0.9	78.3
Australia	50633	0.8	84.4	30142	0.7	79.0
Argentina	9059	0.1	84.5	15992	0.4	79.4
Tunisia	7006	0.1	84.6	18291	0.4	79.8
Other	996611	15.4	100.0	873096.5	20.2	100.0
World	6484979.5	100.0	100.0	4316655	100.0	100.0

Source: Ellaborated on the basis of data from Buckley (1990: 21).

Table 7.3
Imports of Fresh Fruit and Wine and Country Share of World Imports, 1987–1988

	FRESH FRUITS AND VEGETABLES					
	Value (in $ Thousands)	World Share	Cumulative Share	Volume (in Metric Tons)	World Share	Cumulative Share
Germany	4,134,257	18.7	18.7	6,456,445	15.6	15.6
France	2,462,826	11.1	29.8	3,639,965	8.8	24.3
United K.	2,258,702	10.2	40.1	3,386,166	8.2	32.5
United States	1,960,981	8.9	48.9	5,526,639	13.3	45.8
Japan	1,154,195	5.2	54.1	1,534,002	3.7	49.5
Canada	1,249,914	5.7	59.8	2,230,106	5.4	54.9
Netherlands	1,036,011	4.7	64.5	2,438,915	5.9	60.8
Italy	788,108	3.6	68.0	1,702,073	4.1	64.9
Belg./Luxe.	733,514	3.3	71.4	1,737,845	4.2	69.0
Switzerland	593,075	2.7	74.0	516,138	1.2	70.3
Hong Kong	397,301	1.8	75.8	614,767	1.5	71.8
Sweden	493,364	2.2	78.1	682,311	1.6	73.4
Austria	389,241	1.8	79.8	653,240	1.6	75.0
Saudi Arabia	278,960	1.3	81.1	895,600	2.2	77.1
Finland	262,206	1.2	82.3	351,737	0.8	78.0
Other	5,737,966	26.0	108.2	12,333,126	29.7	107.7
Total World	22,109,547	100.0	100.0	41,501,418	100.0	100.0

125

Table 7.3 Continued

	Value (in $ Thousands).	World Share	Cumulative Share	WINE Volume (in Metric Tons).	World Share	Cumulative Share
W.Germany	1,056,341	15.7	15.7	888,672	20.8	20.8
United K.	1,256,011	18.7	34.4	609,892	14.3	35.1
France	303,612	4.5	38.9	484,769	11.4	46.5
United States	942,551	14.0	53.0	299,904	7.0	53.5
Netherlands	377,325	5.6	58.6	209,525	4.9	58.5
Belg./Luxem	427,238	6.4	64.9	205,509	4.8	63.3
E. Germany	150,000	2.2	67.2	195,675	4.6	67.9
Switzerland	407,982	6.1	73.2	204,926	4.8	72.7
USSR	320,989	4.8	78.0	165,932	3.9	76.6
Canada	205,803	3.1	81.1	134,731	3.2	79.7
Other	1,272,101	18.9	100.0	864,349	20.3	100.0
World	6,719,951	100.0	100.0	4,263,882	100.0	100.0

Source: Ellaborated on the basis of data from Buckley (1990: 31, 42).

classification previously presented in Korzeniewicz and Martin (1993) to identify core, semiperipheral, and peripheral nations in our data; the coding in this
instance, at least, is rather standard.[19]

Second, following the procedures outlined in Korzeniewicz and Martin
(1993), we have used the classification of nations according to world-economic
zones to establish the shares of exports and imports accounted for by core,
semiperipheral, and peripheral areas of the world-economy for each of our two
sets of commodities.[20] We then use two indexes to analyze the zonal distribution
of these exports and imports. The "coreness" index measures the extent to
which exports/imports of a given set of commodities is "core" or "peripheral";
we constructed it by dividing core exports/imports by the sum of core and
peripheral exports/imports. According to this index, a commodity that is ex-
ported/imported almost exclusively by firms in the core would approximate a
value of 1.00, while a commodity that is exported/imported almost exclusively
by firms in the periphery would approximate a value of zero.

Our second index is designed to evaluate the extent to which the ex-
ports/imports of our two sets of commodities are located in the semiperiphery
of the world-economy. For each of the two sets of commodities, we have built
a "semiperipherality" index by dividing semiperipheral exports/imports by total
exports/imports. According to this index, a commodity that is heavily ex-
ported/imported by firms in the semiperiphery should approximate a value of
1.00, as opposed to a value of zero when the semiperiphery accounts for little
of overall exports/imports.

Combined, these two indexes allow for a more systematic evaluation of the
zonal distribution of production processes. For example, the indexes can be used
to examine the hierarchies of production processes and commodities that are
explicitly or implicitly built into studies focusing on development and/or the
global division of labor. Figure 7.1 uses these indexes to depict the spatial
location of our two sets of commodities.

As suggested by Figure 7.1 exports and imports of both sets of commodities
are heavily concentrated in core areas of the world-economy. Furthermore, the
data suggest that the overall spatial location of exports (as measured by our two
indexes) for fresh fruit and vegetables underwent little change through the
1980s. The spatial location of wine exports, on the other hand, shifted even
more toward the core, reflecting primarily the drastic decline of Algerian exports
during the 1980s.

The data also suggest some differences between the two sets of commodities
in the spatial location of imports (and presumably consumption of these exports).
In the case of fresh fruit and vegetables, the overwhelming destination of exports
consists of core areas of the world-economy. In the case of wine, semiperipheral
areas (primarily in Latin America and Eastern Europe) account for a considerably larger share of imports.

Of course, this does not mean that these sets of commodities are characterized
by homogeneous returns. Like clothing (see Gereffi 1993), FFV and wine are

Figure 7.1
World-System Location of Exports (E) and Imports (I) of Fresh Fruit and Wine, 1980s

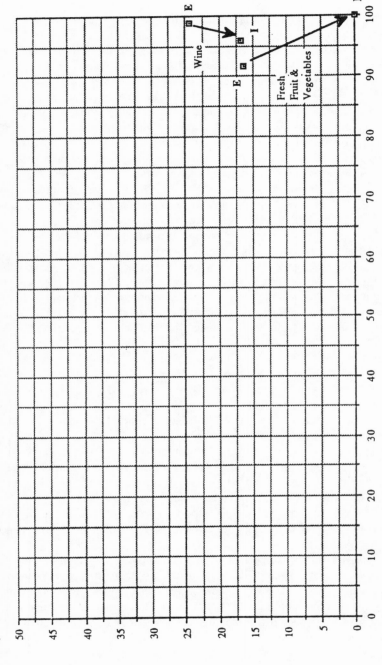

Table 7.4
Average Value of FFV and Wine Exports, 1987–1988

FFV, Av. Dollars per Ton			Wine, Av. Dollars per Metric Ton	
Italy	657.03		France	2530.39
United States	587.70		Australia	1679.82
Spain	557.24		United States	1384.31
France	542.31		Spain	1143.50
Netherlands	522.64		Italy	824.63
Israel	490.07		*Av. Core*	*1632.23*
Canada	109.91			
Av. Core	*526.00*		Portugal	2073.70
			Bulgaria	1050.58
Chile	654.96		Argentina	566.47
South Africa	468.59		Algeria	480.99
Greece	464.88		*Av. SP*	*1366.06*
Mexico	336.94			
Costa Rica	249.33		Tunisia	383.03
Av. SP	*407.89*		*Av. Periphery*	*383.03*
Morocco	380.20		Other	1141.47
Honduras	353.85			
China	251.19		World	1502.32
Av. Periphery	*320.91*			
Other	328.81			
Total World	424.93			
Source: Ellaborated on the basis of data in Buckley (1990, pp. 21, 41).				

stratified commodities, with the higher value or "premium" varieties more often produced in the higher ranking zones of the world-system. A simple indication of existing differences can be developed by calculating the average value of exports from core, semiperipheral, and peripheral areas of the world-economy. As indicated by the data in Table 7.4, the average values for fresh fruit and vegetable exports of core nations are 29.0 percent higher than those of semiperipheral nations, and the average value of exports from semiperipheral nations are 27.1 percent higher than those of peripheral nations. Likewise, in the case of wine production, the average values for exports of core nations are 19.5 percent higher than those of semiperipheral nations, and the average value of exports from semiperipheral nations are 256.6 percent higher than those of peripheral nations (although in this last case, the sample is extremely small).

Another way of highlighting these differences is by discriminating among

Figure 7.2
World-System Location of Exports (E) and Imports (I) of Apples, Bananas, and Grapes, 1989–1990

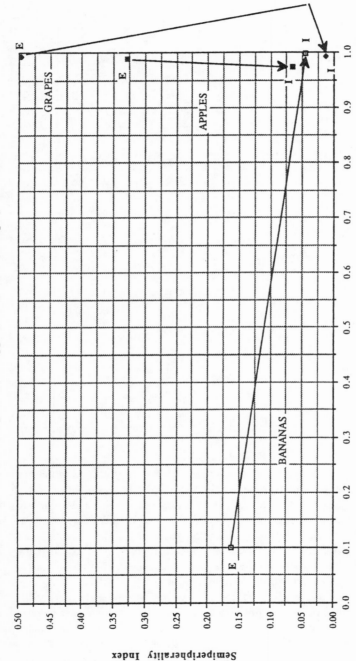

Coreness Index

Source: Elaborated on the basis of data from International Trade Statistics Yearbook (1990).

specific commodities. Using the same indexes as Figure 7.1, Figure 7.2 represents the spatial location of exports and imports of bananas, apples, and grapes. As suggested by the figure, there are significant differences among these commodities. Banana exports and imports would most closely follow the traditional pattern of production in peripheral areas for consumption in the core. Apples and fresh grapes, on the other hand, although consumed primarily in the core, are produced to a greater extent in the semiperiphery of the world-economy, with less of these activities taking place in the core.

Although these observations provide a descriptive visualization of the spatial location of these activities in the world-economy, we can also use them to formulate three hypotheses. First, activities scoring lower on the "coreness" index (such as banana exports) are likely to be controlled more closely by multinational enterprises than activities scoring high in the same index (such as fresh grape or apple exports). This would explain the relative importance of local enterprises in Argentina and Chile. Second, activities scoring lower on the "coreness" index are also likely to be controlled by fewer firms than activities scoring high on the same index. Again, the characteristics of Argentina and Chile would fit these expectations. Third, activities scoring high on the "coreness" index are likely to be characterized by intense processes of innovation that serve to transfer competitive pressures elsewhere in the world-economy. In this sense, the divergent pattern of export strategies in Argentina and Chile would constitute a relational process, in which the very success of enterprises in Chile would reside in intensifying competition within the commodity chains at hand.

We should observe two significant distinctions between the FFVGCC and the WGCC. First, the timing of these chains differs. While the WGCC has been important in the world market for a long time, the expansion of the FFVGCC has been very recent. This characteristic may have additional consequences for the organizational patterns that characterize each GCC. Having been created earlier, the niches in the WGCC are probably more entrenched and more difficult for newcomers to enter. The recent expansion of the FFVGCC, on the other hand, may provide greater opportunities for new enterprises to capture and sustain competitive positions within the chain.

Second, while efforts made by enterprises in the semiperiphery to increase wine exports in the world market enter into direct competition with core producers, many of the principal semiperipheral nations involved in the FFVGCC (e.g., Chile) seasonally complement core production. This may be an additional reason for enterprises in the FFVGCC to find greater flexibility in capturing competitive niches within their chain.

CONCLUSIONS

Theories of development are currently in crisis. Having built the concept of industrialization in their analytical models as the central gear driving the accumulation of wealth, social equity, and democratization across nations, most theories of development (in both their functionalist and critical variations) failed to anticipate the declining significance of manufacturing as an engine of economic growth. World-systems theory is better equipped to provide a coherent and verifiable analysis of these transformations, for it has always focused on global processes of competition and innovation—rather than industrialization—as essential dynamics shaping the accumulation and distribution of wealth in the world-economy. In this chapter, we have briefly elaborated these arguments by focusing on two commodity chains that are often perceived as stereotypical examples of traditional or peripheral economic activities: the fresh fruit and vegetable global commodity chain and the wine global commodity chain in Argentina and Chile.[21]

At all points in the commodity chains involved, states have played an important role in regulating markets and promoting the expansion of some activities and not others. State policy makers design pertinent strategies in pursuit of their own objectives (such as increasing state revenues, reducing unemployment, or alleviating social tensions). In this sense, state policies have had an important impact in shaping competitive opportunities and constraints within commodity chains. Comparative advantages and markets in both core and semiperipheral areas, we suggest, are socially and politically constructed.

The characteristics of the labor force have also been a significant variable shaping these commodity chains. But low labor costs are not enough of an explanation for the comparative success of enterprises in Chile. Insofar as both fresh fruits and wines can be considered as part of GCCs, and to the extent that exports of these commodities tend to be heavily centered in the core, low wages are not a universal characteristic of the competitive advantages that provide for success in these activities. Other variables must be examined as well.

Finally, innovation appears as a crucial component of competitive advantages in these commodity chains. As activities centered in core and semiperipheral nations, these commodity chains are characterized by continuous change in their organization. Successful export strategies have been rooted in an enhanced ability to maintain and capture competitive niches through constant innovation.

But even successful strategies of growth (as pursued by state agencies or business enterprises) tend to generate new constraints over the long run. It is in facing these constraints that the comparative advantages of core location for business enterprises become most apparent. The first important constraint is in regard to labor: The very success of these strategies tends to create conditions (tight labor markets, growing demand for raw materials) that strengthen the bargaining power of workers. Second, successful models of development invite imitation by other participants (e.g., states, enterprises) in the world-economy. In the process, the competitive edge provided by these models of development tends to disappear. In Chile, these two constraints are likely to become increasingly apparent in the immediate future.

NOTES

This project was made possible in part by research grants to Miguel E. Korzeniewicz and Roberto P. Korzeniewicz by the North-South Center of the University of Miami, and by the Heinz Foundation through the Center for Latin American Studies of the University of Pittsburgh. Victoria Carty provided invaluable research assistance for this project. We would like to thank Phillip McMichael for his comments on a previous version of this chapter.

1. For descriptive comparisons, see Gil and Sansavini (1990) and Bergougnoux and Vidaud (1990).

2. For Rivera (1991, 2), "the institutional framework established between the state and the private sector . . . is one of the principal factors that explains the Chilean fruit export expansion."

3. This type of situation is not limited to this particular case. In his study of garment production for export in Colombia, Morawetz (1981, 31) indicated that "all thirteen exporters to the United States and Europe who gave us price comparisons indicated that the price received on export sales (including the value of export incentives) was below the domestic price."

4. In Argentina, these actions were portrayed as a consequence of Chilean fears over competition, disguised as sanitary concerns (*La Nacion* [Argentina], May 30, 1992, sec. 3, p. 1). Competitive concerns were at least part of the equation: Jorge Prado, president of the Sociedad Nacional de Agricultura in Chile, indicated that "some pretend to give away to our competitors the use of the port and road infrastructure of the country" and challenged "this giveaway of one of the most important comparative advantages that our national agriculture owns" (*El Mercurio* [Chile; Ed. Internacional], April 9–15, 1991). Lower transportation costs were certainly an important advantage facilitating the early development of fruit exports in Chile (Cruz and Leiva 1982, 7), and the state subsidized major improvements in highways and port facilities.

5. In Chile, Cruz and Leiva (1982, 135–65) indicated that labor costs accounted for 42.3 percent of all costs for small firms, 46 percent of costs for medium enterprises, and from 43 to 48 percent for large enterprises (for the large enterprises, 80 percent of labor costs were accounted for by seasonal workers).

6. Sáez (1986, 12) has emphasized the impact after 1973 of "the constitution—for the first time in the history of Chile—of an open market in land and capital that reordered the economic, social and political organization of agriculture." The open market now also involved labor (Cruz and Leiva 1982, 113).

7. The participation of women is shaped by the very timing of seasonal work. The summer months allow for greater flexibility in women's schedules, particularly insofar as some of them are students and others are responsible for the care of children and their preparation for school.

8. Attitudes toward working conditions may be related to the length of participation in the labor force. Medel R., Olivos M., and Riquelme G. (1989, 38–39) suggest that women with longer experience as seasonal workers tended to be more critical of their employers.

9. Wages in Chile have also been shaped by labor unrest, notably brief work stoppages during peak periods in the packing of fresh fruit (Medel R., Olivos M., and Ri-

quelme G. 1989, 93). Falabella (1988, 2) indicates that these stoppages revolve around working conditions and tend to be organized spontaneously.

10. A trade article regarding fresh fruit exports indicated that "all our activity is labor-intensive, and the current economic program foresees the discussion of wages in open negotiations. This is shaping up as a factor that will increase costs, given the inelasticity of the supply of labor in the sparsely populated region where we must develop our activities. Thus, average yearly wages have increased above the levels of past years, which undoubtedly must be charged to the costs of production" (Barzi 1991, 58–59).

11. Examples of these innovations abound. In the future, according to industry analysts, "fruits and vegetables will be produced with far fewer chemicals because disease and pest resistance will be bred or bioengineered into them" (*The Packer*, December 28, 1991, p. Ia). Conveyor belts using water to move fruit have been adopted to reduce product damage, as "equipment that handles product more gently is in greater demand as consumers look for blemish-free produce" (*The Packer*, December 7, 1991, p. 3c). Computers and optical devices have been incorporated into production to grade, weigh, and classify fruit with greater precision and speed, saving on final costs.

12. For example, the installation of controlled atmosphere technology with a one-million-box storage capacity by Standard Trading at Teno, Chile, in 1985 required a $33 million investment (Sáez 1986, 16).

13. New planting methods were limited to large producers who could afford the investment and the long period of transition between old and new production (Manzanal and Rofman 1990, 128).

14. The same characteristics apply to industry concentration and foreign participation in these commodity chains. Despite claims to the contrary, industry concentration has not become more pronounced. In the case of Chile's overall fruit exports, the largest five firms controlled 47.4 percent of exports in 1983, and 44.2 percent in 1991. Likewise, for the ten largest firms, the share of exports actually declined from 64.7 percent in 1983 to 58.7 percent by 1991. If we were to focus more specifically on the case of grapes and apples, the share of the largest five firms declined from 72.6 percent of all exports in 1979, to 49.2 percent in 1983, and 48.5 percent by 1991. For the ten largest firms, the share of exports showed a similar rate of decline, from 86.5 percent in 1979, to 72.5 percent in 1983, and 65.3 percent by 1991 (elaborated on the basis of data from Asociacion de Exportadores de Chile 1992). Likewise, although many have argued that fresh fruit production in Chile has been increasingly taken over by multinationals, the evidence does not fully corroborate this interpretation. In Chile, foreign firms moved into fresh fruit production in the early 1980s, and quickly controlled roughly a quarter of all fresh fruit exports by 1984. But the growth in the share of the foreign firms after the initial shift of production that followed the 1983 crisis has been rather slow: from 23.6 percent of all fresh fruit exports in 1984–1985, their share had increased to 30.5 percent by 1991–1992 (elaborated on the basis of data from Abrahams and Contreras 1986, 97–98; Asociacion de Exportadores de Chile 1992). Although these data are not conclusive, they clearly suggest that greater research is needed to concentration and foreign participation in these commodity chains.

15. In the late 1980s, 14.2 percent of the labor force in Rio Negro was employed in the fresh fruit sector: 6.4 percent in primary activities, 5.8 percent in packing, and 2 percent in industrial processing (Castello, Pacenza de Del Franco, and Izurieta 1990, 216).

16. The data are from Buckley (1990). The data are for the fifteen top exporters and

importers. These top fifteen nations accounted in 1987–1988 for 71.3 percent of world exports of fresh fruits and vegetables.

17. This is important to emphasize: Friedland (1991, 8), for example, neglects to highlight the importance of core areas in the production and export of fresh fruits and vegetables.

18. We are forced to use imports as a rough indicator. Except for the OECD nations, few data are available on worldwide fresh fruit consumption (see Buckley 1990, 14). As indicated by Friedland (1991, 8), large countries (such as China, India, and Nigeria) have "only local production for local consumption."

19. For example, among the fifteen top exporting nations in fresh fruit and vegetables listed in Table 7.2, core nations are Spain, the Netherlands, Italy, the United States, France, Canada, and Israel; semiperipheral nations are Mexico, Chile, South Africa, Greece, and Costa Rica; and peripheral nations are Honduras, China, and Morocco.

20. The following section borrows from Korzeniewicz and Martin (1993).

21. Hopkins and Wallerstein (1986, 159) have defined a commodity chain as "a network of labor and production processes whose end result is a finished commodity." As indicated elsewhere, a "GCC consists of sets of inter-organizational networks clustered around one commodity or product, linking households, enterprises and states to one another within the world-economy. These networks are situationally specific, socially constructed, and locally integrated, underscoring the social embeddedness of economic organization. . . . Specific processes or segments within a commodity chain can be represented as boxes or nodes, linked together in networks. Each successive node within a commodity chain involves the acquisition and/or organization of inputs (e.g., raw materials or semifinished products), labor power (and its provisioning), transportation, distribution (via markets or transfers), and consumption. The analysis of a commodity chain shows how production, distribution and consumption are shaped by the social relations (including organizations) that characterize the sequential stages of input acquisition, manufacturing, distribution, marketing and consumption" (Gereffi, Korzeniewicz, and Korzeniewicz 1993, 32).

REFERENCES

Abrahams, Joyce, and Patricia Contreras. 1986. "La comercialización de la fruta fresca de exportación." *Agricultura y Sociedad* (Santiago de Chile) 3 (July): 75–99.

Arrigli, Giovanni, and Jessica Drangel. 1986. "The Stratification of the World-Economy: An Exploration of the Semi-Peripheral Zone." *Review* 10, 1: 9–74.

Asociacion de Exportadores de Chile. 1992. *Informe estadístico de exportación de frutas y hortalizas.* Santiago de Chile: Asociación de Exportadores de Chile.

Barzi, Guillerrmo A. 1991. "Alto Valle del Rio Negro: Ante un nuevo panorama." *Anales de la Sociedad Rural Argentina* (Argentina) 124, 4–6 (April–June): 57–61.

Bergougnoux, François, and Jacques Vidaud. 1990. "La produzione di uve da tavola in Cile e Argentina." *Rivista di Frutticoltura* (Italy) 52, 1 (Gennaio): 65–68.

Buckley, Katharine C. 1990. *The World Market in Fresh Fruit and Vegetables, Wine and Tropical Beverages—Government Intervention and Multilateral Policy Reform.* Staff Report No. AGES 9057. Washington, D.C.: U.S. Department of Agriculture.

Castello, Hector L., Irma Beatriz Pacenza de Del Franco, and Carlos Izurieta. 1990. "La actividad frutícola en el Alto Valle de Rio Negro.'' In *Agroindustrias en la Argentina: Cambios organizativos y productivos (1970–1990)*, edited by Graciela Gutman and Francisco Gatto, pp. 215–46. Buenos Aires: Centro Editor de América Latina.

Collins, Jane L. 1991. ''Production Relations in Irrigated Agriculture: Fruits and Vegetables in the São Francisco Valley (Pemambuco/Bahia, Brazil).'' Working paper no. 23 of the Workshop on the Globalization of the Fresh Fruit and Vegetable System, University of California at Santa Cruz, December 6–9.

Comisión Económica para América Latina y el Caribe. 1991. ''Los canales de comercialización y la competitividad de las exportaciones latinoamericanas.'' Document prepared for the seminario sobre los canales de comercialización y la competitividad de las exportaciones latinoamericanas, Santiago de Chile, September 9–12.

Cruz, M. Elena, and Cecilia Leiva. 1982. *La fluricultura en Chile después de 1973: Un area privilegiada del capital*. Santiago de Chile: Grupo de Investigaciones Agrarias.

De Luca, Marta. 1991. ''Proyección de la industria vitivinícola en el Mercado Común del Sur.'' Paper presented at the Seminario Internacional de la Vitivinicultura, San Rafael (Mendoza), Argentina, November 10–15.

Falabella, Gonzalo. 1988. ''El sistema de trabajo temporal (o la institucionalización de la desconfianza, la incertidumbre y la desorganización social).'' Paper presented at the Seminario Internacional sobre la Agricultura Latinoamericana: Crisis, Transformaciones y Perspectivas, Punta de Tralca, Chile, September 1–4.

Friedland, William H. 1991. ''The Global Fresh Fruit and Vegetable System: An Industrial Organizational Analysis.'' Working paper no. 4 of the Workshop on the Globalization of the Fresh Fruit and Vegetable System, University of California at Santa Cruz, December 6–9.

Gabriel y Cia., S.R.L. 1992. *Estadística de la exportación de frutas de la República Argentina al 31/12/91*. Buenos Aires: Gabriel y Cia., S.R.L.

Gereffi, Gary. 1993. ''The Role of Big Buyers in Commodity Chains: How U.S. Retail Networks Affect Overseas Production Patterns.'' In *Commodity Chains and Global Capitalism*, edited by G. Gereffi and M. E. Korzeniewicz, 95–122. Westport, Conn.: Greenwood Press.

Gereffi, Gary, Miguel Korzeniewicz, and Roberto P. Korzeniewicz. 1993. ''Introduction: Global Commodity Chains.'' In *Commodity Chains and Global Capitalism*, edited by G. Gereffi, and M. E. Korzeniewicz, 1–14. Westport, Conn.: Greenwood Press.

Gil, Gonzalo, and Silviero Sansavini. 1990. ''Tendenze produttive e complementarietá fra frutticoltura sudamericana ed europea: Riflessi sulla situazione italiana.'' *Rivista de Frutticoltura* (Italy) 52, 1 (Gennaio): 13–21.

Gleazer, E. Allen. 1990. ''Wineries Get with the Program.'' *Wines and Vines* (USA) 71, 3 (March): 19–22.

Goldfrank, Walter L. 1991. ''Chilean Fruit: The Maturation Process.'' Working paper no. 16 of the Workshop on the Globalization of the Fresh Fruit and Vegetable System, University of California at Santa Cruz, December 6–9.

———. 1993. ''Fresh Demand: The Consumption of Chilean Produce in the USA.'' In *Commodity Chains and Global Capitalism*, edited by G. Gereffi and M. E. Korzeniewicz, 267–79. Westport, Conn.: Greenwood Press.

Gomez, Sergio. 1991. ''La uva chilena en el mercado de los Estados Unidos.'' Working paper no. 18 of the Workshop on the Globalization of the Fresh Fruit and Vegetable System, University of California at Santa Cruz, December 6–9.

Gomez, Sergio, and Jorge Echenique. 1991. *La agricultura chilena: Las dos caras de la modernización.* Third edition. Santiago de Chile: FLACSO.

Hopkins, Terence K., and Immanuel Wallerstein. 1986. "Commodity Chains in the World-Economy Prior to 1800." *Review* 10, 1: 157–70.

How, R. Brian. 1991. *Marketing Fresh Fruits and Vegetables.* New York: Van Nostrand Reinhold.

Instituto Nacional de Vitivinicultura (Mendoza, Argentina). 1991. *Exportaciones argentinas de productos vitivinícolas, 1990.* Mendoza, Argentina: Instituto National de Vitivinicultura.

———. 1992. *Exportaciones argentinas de productos vitivinícolas. 1990. (Anticipo).* Mendoza, Argentina: Instituto Nacional de Vitivinicultura.

Juri, Maria de la Esperanza. 1991. *Exportaciones argentinas de productos perecederos: Evolución, problemas y perspectivas. Una comparación con el caso chileno.* Santiago de Chile: Oficina Regional de la FAO para América Latina y el Caribe.

Korzeniewicz, Roberto, and William Martin. 1993. "The Global Distribution of Commodity Chains." In *Commodity Chains and Global Capitalism*, edited by G. Gereffi and M. E. Korzeniewicz, 67–91. Westport, Conn.: Greenwood Press.

Lieber, Robert M. 1990. "The Changing Labor Picture in the Vineyards for the 1990s." *Wines and Vines* (USA) 71, 2 (February): 34–35.

Llambi, Luis. 1991. "The Political Economy of Non-Traditional Fruit and Vegetable Exports from Latin America." Working paper no. 15 of the Workshop on the Globalization of the Fresh Fruit and Vegetable System, University of California at Santa Cruz, December 6–9.

Manzanal, Mabel, and Alejandro B. Rofman. 1990. *Las economías regionales de la Argentina. Crisis y políticas de desarrollo.* Buenos Aires: Centro Editor de América Latina.

Medel R., Julia, Soledad Olivos M., and Verónica Riquelme G. 1989. *Las temporeras y su visión del trabajo (condiciones de trabajo v participación social).* Santiago de Chile: Centro de Estudios de la Mujer.

Mendez Boaglia, Maria V., and José Luis Bonifacio. 1992. "El trabajo frutícola en el Alto Valle de Rio Negro en Neuquén." Paper presented at the Primer Congreso Nacional de Estudios del Trabajo, Buenos Aires, Argentina, May 26–29.

Morawetz, David. 1981. *Why the Emperor's New Clothes Are Not Made in Colombia.* New York: Oxford University Press.

Muñoz Honorato, Ivan. 1991. "Uvas de mesa, producción, sistema de comercialización en la vitivinicultura chilena." Paper presented at the Seminario Internacional de la Vitivinicultura, San Rafael (Mendoza), Argentina, November 10–15.

Ominami, Carlos, and Roberto Madrid. 1988. *La inserción de Chile en los mercados mundiales.* Santiago de Chile: Dos Mundos.

Organización de las Naciones Unidas para la Agricultura y la Alimentación. 1985. "La economía del vino en América Latina. Primera parte." *Comercio Exterior* (Mexico) 35, 5 (May): 506–15.

———. 1985. "La economía del vino en América Latina. Segunda parte." *Comercio Exterior* (Mexico) 35, 6 (June): 612–22.

Quintana, Martha Sofia. 1992. "Como con bronca y junando." *Frutos* (Argentina) 9 (January): 30–31.

Rivera, Rigoberto. 1991. "Institutional Conditions of Chilean Fruit Export Expansion."
Working paper no. 17 of the Workshop on the Globalization of the Fresh Fruit
and Vegetable System, University of California at Santa Cruz, December 6–9.
Sáez, Arturo. 1986. "Uvas y manzanas, democracia y autoritarismo: El empresario fru-
tícola chileno (1973–1985)." In *Documento de Trabajo del Grupo de Investi-
gaciones Agrarias*. Santiago de Chile: Academia de Humanismo Cristiano.
Scoblionkov, Deborah. 1992. "Argentina's Surprise Hit—Thanks to an American." *Phil-
adelphia Inquirer* (USA), November, M1, M7.
United Nations, Department of Economic and Social Affairs. 1978–1990. *International
Trade Statistics Yearbook*. New York: United Nations.
Walker, Larry. 1987. "A Wines and Vines Focus on Barrels." *Wines and Vines* (USA)
68, 10 (October): 16–20.
Wilkinson, John. 1991. "A Profile of the Brazilian Fruit and Vegetable for Export
Sector." Working paper no. 24 of the Workshop on the Globalization of the Fresh
Fruit and Vegetable System, University of California at Santa Cruz, December
6–9.

8

The Regulation of the World Coffee Market: Tropical Commodities and the Limits of Globalization

John M. Talbot

Recent work within the world-system paradigm has greatly increased our understanding of the organization of world markets and of the ways in which world trade has incorporated peripheral and semiperipheral countries into a globally organized economy. Food regime theory, originally developed by Harriet Friedmann, has analyzed the increasing industrialization and globalization of food production and trade. Commodity chain analysis, developed by Gary Gereffi and Miguel Korzeniewicz (1990), has focused attention on the linked sequences of processing stages that connect the production of raw materials to the sale of finished products to consumers, which often cross national boundaries. In this chapter, I will argue that there is a class of tropical commodities, here represented by coffee, that has been overlooked by both of these analytical schools. Insights drawn from both schools help us to analyze the organization of the world coffee market, but a focus on coffee as the archetypal tropical commodity reveals two aspects of the organization of international trade that have not been fully incorporated into either school. First, many of the poorest peripheral countries are still linked to the world economy primarily as producers and exporters of tropical commodities, a pattern established during the colonial period which persists today despite the increasing globalization of agricultural production. Second, the nature of tropical commodity production, trade, and dependence has led Third World producing states to attempt to overcome their disadvantageous position in the world economy through collective actions to change the political structures of world markets for these commodities.

The recent work of Friedmann and McMichael (Friedmann 1991a, 1991b; McMichael and Myhre 1991; McMichael 1992; Friedmann and McMichael 1989) has identified and analyzed two distinct food regimes that have governed international agricultural production and trade. The first, established during the imperialist period (ca. 1880–1914), was characterized by the rapid increase of food production in the settler states, and by food exports from these states which competed directly with European agriculture. This new form of agricultural trade between nation-states supplanted the traditional colonial trade in tropical agricultural products as the most important dynamic shaping the world food system. The first regime broke down during the economic crisis following World War I, but the U.S. model of agriculture as an integrated sector of a national industrial-capitalist economy formed the basis for the second food regime, which was established under U.S. hegemony following World War II. This second regime was based on regulation of national economies by independent nation-states (Fordism in the United States and Western Europe and state-led industrialization in the Third World), but it was increasingly undercut by the rapid expansion of transnational corporations (TNCs) as the organizers of global production and trade.

Friedmann identifies three major characteristics of this second regime: (1) the *wheat complex*, involving agricultural subsidy programs in the core, which led to overproduction of wheat and other grains, and rapidly expanding grain exports to the Third World, led by U.S. "food aid" (Friedmann 1978, 1982); (2) the *durable food complex*, involving import substitution by TNCs of ingredients produced in the core for traditional tropical imports from the Third World (e.g., beet sugar and high fructose corn syrup for cane sugar) for use in highly processed foods sold to core consumers; and (3) the *livestock/feed complex*, involving transnational integration and industrialization of the commodity chains leading from agricultural products through livestock feed, livestock, and finally to meat products sold in core markets and to the rich and middle classes in the Third World.

The second food regime was thrown into crisis by the combined energy-food-currency shocks of the early 1970s. The three tendencies toward transnationalization of food production outlined above had undermined the possibility of national regulation of agricultural production. The debt crisis and World Bank and International Monetary Fund policies of structural adjustment and export promotion began to force Third World producers, whose traditional exports had been undercut by transnationalization, into new specializations in nontraditional agricultural exports, also organized by the TNCs. Friedmann and McMichael identify several tendencies at work in the struggle to create a new globally organized food regime: (1) the Uruguay Round of the General Agreement on Tariffs and Trade (GATT) negotiations, in which the United States is seeking to extend free trade to previously excluded agricultural markets, to enable U.S.–based agro-industrial TNCs to gain access to protected markets in Europe and Japan; (2) the emergence of the newly agricultural countries (NACs, a term coined by Friedmann), semiperipheral industrial-agricultural exporters who could compete with the United States and Europe on increasingly liberalized world agricultural markets, leading to trade wars; and (3) the introduction of biotechnology, increasing tendencies toward *substitutionism*, the diversification and standardization of agricultural inputs into the industrial food system of the TNCs, and *appropriationism*, the introduction of new, industrially produced inputs into the agricultural system, making all agricultural producers more heavily dependent on ''proprietary seed-chemical packages,'' controlled by the TNCs (Goodman 1991).

This literature provides valuable insights into the reorganization of world agricultural production which is currently under way. But these analyses focus on agricultural products involved in the wheat, durable foods, and livestock/feed complexes, and overlook traditional agricultural products which remain important in world trade. Because of this focus, these analyses also tend to overlook the poorer peripheral countries which are still major exporters of these tropical commodities, many of which are neither newly industrialized countries (NICs) nor NACs. Further, these analyses do not incorporate the possibilities for collective action by Third World agricultural producers to change the political structures of world markets for these commodities. However, the production and trade of tropical commodities have also been affected by the tendencies toward transnationalization, as identified by Friedmann and McMichael. While Third World states had some success in imposing various forms of regulation on tropical commodity markets during the 1960s and 1970s, their efforts also provoked a reaction by the TNCs, allied with core states. During the 1980s, transnationalization and political reaction undermined the possibilities for further collective action by producers. Despite this, coffee producers were able to sustain the political regulation of the world coffee market until 1989, and they still retain some capacity collectively to influence its structure.

In the following sections, the distinctive characteristics of the tropical commodities in general, and of coffee in particular, are described. Second, the importance of tropical commodity exports to poor peripheral countries is documented, as well as the persistence of the colonial trading pattern. Next, I will analyze the ways in which the coffee-producing states acted collectively to restructure the market, during the period of the second food regime, and compare the regulation of the coffee market to that of the food regime. Finally, I will show how the increased globalization of agricultural production since the end of the second food regime has affected the capacity of coffee producers to intervene in the market, and I will attempt to explain why they retained this ability into the 1980s. My argument is that food regime theory overstates the tendencies toward globalization of agricultural systems and underestimates the countertendencies toward Third World state-capitalist alliances which challenge TNC control of agricultural production.

THE TROPICAL COMMODITIES

For a number of traditional agricultural exports which became important in world trade during the colonial era or earlier, the international division of labor established over a century ago persists. This is shown in Table 8.1, which lists

Table 8.1
Patterns of Production and Trade for Major Primary Commodity Exports of the Third World

Commodity	Total Third World Export Earnings*	Third World Share of World Exports	Core Share of World Imports
Petroleum	87,535	68.2%	80.6%
Coffee	9,349	88.3	92.3
Tropical timber	8,488	65.7	76.1
Sugar	7,924	68.2	70.2
Copper	7,715	59.5	79.2
Cotton and cotton yarn	7,012	54.9	64.4
Natural rubber	4,848	97.1	67.8
Iron ore	3,318	44.3	85.3
Soybeans and soybean oil	2,753	32.5	60.5
Cocoa	2,537	95.4	91.4
Palm oil	2,484	95.9	28.3
Rice	2,334	59.7	32.2
Tea	1,954	86.0	54.0
Bananas	1,906	94.4	94.2
Tobacco	1,858	44.9	81.2

Tropical oils and oilseeds	1,633	84.0	80.5
Coarse grains	1,530	12.0	67.2
Beef	1,381	12.2	83.6
Tin	1,142	80.7	73.0
Phosphate rock	1,063	65.0	71.4
Wheat and wheat flour	1,060	6.3	42.2
Bauxite	808	88.1	86.5
Hides, skins, and furs	725	9.5	72.4
Jute and jute products	622	87.2	67.8
Wool	572	9.2	74.4
Hard fibers and products	244	69.5	90.9
Manganese ore	211	45.9	83.5

*millions of US dollars

Source: UNCTAD, *Commodity Yearbook 1990* (New York: UN, 1990). The core here includes USSR and Eastern Europe, except Yugoslavia. China, Cuba, Vietnam, etc. are all included as part of the Third World.

the leading primary commodities exported by Third World producers, ranked in order of the total value of Third World exports. For each commodity, the table shows the total percentage of world exports accounted for by Third World exporters, and the total percentage of world imports taken by core countries. The three complexes identified by Friedmann can be seen in this table. The wheat complex is exemplified by wheat and wheat flour, which are produced mainly in the core and consumed predominantly in the Third World. The effect of the durable food complex can be seen in the case of palm oil, produced in the Third World, but now mainly exported to other Third World countries because it has been substituted for by temperate oils in core markets. The livestock/feed complex is visible in soybeans and soybean oil and coarse grains, two major constituents of feed, and in beef. These products are produced in, and traded primarily among, core countries. Thus, Table 8.1 shows that some of the commodities involved in these three complexes, even if they are produced mainly in the core, and even if their trade is largely controlled by TNCs, remain important sources of income for some Third World producers.

If we shift focus from core consumption, which is a major concern of food regime theory, to Third World production, as shown in Table 8.1, we see a class of commodities with a different pattern of production and trade, which has been considered only briefly by Friedmann and McMichael. This is the class of tropical commodities, produced in the Third World and consumed primarily in the core, which remain very important sources of income for Third World producers. This class includes coffee, tropical timber, sugar, natural rubber, cocoa, tea, bananas, tropical oils and oilseeds,[1] jute, and hard fibers. These commodities have several characteristics in common. Most important, they all require tropical climates for their growth; if consumers in the temperate core countries want to consume them, they must be obtained from the tropical Third World. Production of these commodities is labor intensive and, at least in the early processing stages, not highly mechanized. In addition, no suitable industrially or core-produced substitutes have been found for most of them.[2] Further, major exporters of one of these commodities tend to be major exporters of others as well (e.g., Côte d'Ivoire exports coffee and cocoa, Colombia exports coffee and bananas, and Indonesia exports coffee and tropical timber).

The following analysis focuses primarily on coffee as the most important representative of this class of tropical commodities. Coffee is the most valuable tropical commodity traded on the world market, and it has a greater impact on a larger number of countries than any other. Coffee producers have also been more successful than any other tropical commodity producers in imposing forms of regulation on the world market. Understanding these forms of regulation requires, first of all, an understanding of the coffee commodity chain and of the structure of the coffee production and trading system.

The coffee commodity chain is a relatively simple one, consisting of three main stages between coffee trees in the Third World and consumers in the core. First, growers and exporters perform the processing necessary to produce green coffee beans. This first stage is generally under the control of the producing states, either directly through state marketing boards or indirectly through price, taxation, and exchange rate policies. Second, the green coffee is sold on the world market to importers who are either coffee TNCs or large commodity traders. The international trade is structured by an oligopsony of TNCs on one side (General Foods and Nestlé are the largest) and an oligopoly of producing countries (led by Brazil and Colombia) on the other. The third stage, the final processing of roasted or instant coffee, is dominated by the TNCs, who hold oligopolistic positions in the core markets.

Four major types of coffee are distinguished in the trade. Three of these are arabica coffees, which together make up about 70 percent of total world production. This class is further subdivided into Colombian milds, generally considered to be the highest quality for roasting and brewing; other milds, produced mainly in Latin America, in some African countries, and in Indonesia, which are also considered good for roasting and brewing; and Brazilian milds, which are considered to be of lower quality. The fourth type is robusta, a different variety of coffee bean produced primarily in Africa, which has a harsher taste, but which produces a higher yield of instant coffee per pound than the arabica variety. Most of the major TNC roasters blend coffees of different types in their brand name blends, such as General Foods' Maxwell House. This allows them to adjust the composition of the blend with beans from different sources depending on availability and price, while maintaining a relatively consistent taste.

The structure of the coffee commodity chain is dictated to some extent by the characteristics of coffee. Green coffee can be stored for several years, but roasted coffee retains its flavor only for a few months and must be produced close to the point of consumption. This is why the world trade is conducted mainly in green coffee and why TNCs are able to control access to the core markets for roasted coffee. Instant coffee has a much longer shelf life than roasted coffee and can be produced within the coffee-growing countries and shipped to core markets, but it is highly capital intensive and involves technology that is often controlled by the TNCs.[3] Thus, there are significant barriers to Third World producers extending their control over the coffee commodity chain beyond the green coffee stage; if they want to increase their earnings from coffee, they must change the structure of the world market to raise the price of green coffee.

Another characteristic of coffee creates inherent instability in the world market. Coffee is a tree crop, and there is a lag of from three to five years, depending on the variety, between planting and the time when the trees begin to bear fruit. Yields increase over the next from two to five years and then decline slowly over the next from five to ten years. The major investment in coffee growing is at the planting stage; after that, the main costs are for labor to maintain the trees and harvest the coffee. Because of the biology and economics of coffee growing, supply responds very slowly to price changes, and world coffee prices tend to fluctuate wildly.[4]

THE TROPICAL COMMODITY EXPORTERS

If we focus on the exporters of tropical commodities, we can see the two major characteristics of this trade: the heavy dependence of Third World countries on exports of these commodities and a persistence of trading patterns from the colonial period. Table 8.2 shows the heavy dependence of a number of Third World countries on tropical commodity exports in 1967, near the end of the second food regime period. But it also shows the concentrated nature of market shares in world markets for these commodities. For coffee, the seven largest exporters all depended on coffee for at least 30 percent of their total export earnings, and they controlled 70 percent of world exports; nine additional producers with smaller market shares also depended on coffee for at least 20 percent of their export earnings. Five exporters controlled over 80 percent of world cocoa exports; with the exception of Brazil, the major coffee exporter, these exporters all depended on cocoa for at least 20 percent of their export earnings. Six sugar exporters controlled over 70 percent of world exports, and three of these depended on sugar for more than 50 percent of their export earnings; the others were Brazil, again, and Mexico and the Philippines, large countries dependent on a variety of primary commodity exports. An additional seven countries with small market shares depended on sugar for at least 20 percent of their export earnings. Four tea exporters controlled almost 90 percent of world exports; Sri Lanka and Malawi were heavily dependent on this commodity. Kenya, dependent on coffee for almost 30 percent of its export earnings, had less dependence on tea, and tea was India's single most important export, even though it accounted for only 15.7 percent of total exports. Statistics on market shares were not available for the other tropical commodities considered here, but at least fourteen additional countries depended on one of them for at least 20 percent of export earnings. The major producing states were thus forced to consider tropical commodity export policies as central elements of any economic development strategy, and their concentrated market power gave them the capacity for collective action to increase their export earnings from these commodities.

At the same time, many of these major tropical commodity exporters had, as their primary international trading partner, their former colonial power. Table 8.3 shows that the colonial trading pattern for coffee persists even into the 1990s. Brazil and Colombia are the exceptions. The two largest exporters, they both have strong quasi-state coffee growers' federations with a great degree of control over state coffee policy, which has led to aggressive state-capitalist initiatives to find new markets for their coffees. The destinations of their exports must closely match the distribution of total world imports. The Central American producers tend to export disproportionately to the U.S. market,[5] and the African producers export disproportionately to the old colonial powers in the European Economic Community and to other importers, mainly Asian NICs, the Middle East, and other African coffee importers (see Table 8.3). Indonesia, the major

Table 8.2
Export Dependence and Market Concentration for Four Major Tropical Commodities, 1967

COFFEE

Exporter	Market Share	Dependence on Coffee
Brazil	33.2	44.3
Colombia	14.6	63.2
Angola	5.6	51.9
Cote d'Ivoire	4.8	32.5
El Salvador	4.5	47.7
Uganda	4.4	53.5
Guatemala	3.2	35.2
Total	70.3	

COCOA

Exporter	Market Share	Dependence on Cocoa
Ghana	31.6	73.5
Nigeria	24.5	23.1
Brazil	13.4	5.1
Cote d'Ivoire	10.5	20.4
Cameroon	6.8	30.7
Ecuador	3.9	12.6
Total	90.7	

Table 8.2 Continued

SUGAR

Exporter	Market Share	Dependence on Sugar
Cuba	39.9	83.2
Philippines	10.3	18.9
Dominican Republic	6.1	57.7
Mexico	5.9	8.5
Brazil	5.7	5.0
Mauritius	4.1	94.5
Total	72.0	

TEA

Exporter	Market Share	Dependence on Tea
India	45.1	15.7
Sri Lanka	38.9	65.1
Kenya	3.7	13.9
Malawi	2.2	27.2
Total	89.9	

Source: UNCTAD, *Handbook of International Trade and Development Statistics, 1976* (New York: UN, 1976), Table 4.4.

Table 8.3
Destination of Coffee Exports from Major Coffee Producers, 1990

			Importer		
Exporter	US and Canada	EEC	Other Europe	Japan	Other[4]
Share of Total World Imports[2]	22.1%	46.5%	10.2%	6.9%	14.3%
Brazil[1]	28.8	35.0	13.3	8.0	14.9
Colombia[1]	21.2	50.5	9.3	7.9	11.1
Mexico[1]	85.3	8.0	2.8	1.4	2.5
Guatemala[2,3]	63.6	26.5	8.7	6.0	0.0
El Salvador[2]	30.6	38.4	1.8	1.7	27.5
Costa Rica[2]	21.8	43.5	18.6	3.8	12.3
Honduras[2]	31.2	32.7	14.0	17.5	4.6

Table 8.3 Continued

Cote d'Ivoire[2]	3.5	39.0	1.1	0.8	55.6
Kenya[2]	5.6	43.9	14.6	0.4	35.5
Uganda[2]	7.7	49.6	0.9	2.1	39.7
Zaire[2]	0.7	51.4	2.6	0.0	45.3
Ethiopia[2]	11.6	31.4	3.8	12.8	40.4
Tanzania[2]	1.0	48.2	4.4	8.3	38.1
Indonesia[1]	11.2	47.6	1.2	18.4	21.6
Papua New Guinea[2]	2.9	52.0	2.5	1.9	40.7

Source: UN, *International Trade Statistics Yearbook 1990* (New York, 1992), Trade Matrix Tables. These figures are for SITC 071, which includes instant and other processed forms of coffee.

[1] As reported by the exporting country.

[2] As reported by the importing country.

[3] Due to different time periods used in reporting by the importing and exporting countries, figures add to more than 100%.

[4] This table includes only the 18 largest coffee importers. Other large importers, which account for an additional 9.4% of imports, are (in order of total value of imports): Algeria, Australia, Korea, Portugal, Singapore, Argentina, Saudi Arabia, Israel, Andorra, Morocco, Hong Kong, Ireland, South Africa, Lebanon, Sudan, Malaysia, New Zealand, Jordan, Senegal, Kuwait, Turkey, Egypt.

Asian producer, exports both to Europe, the old colonial power, and to Japan, the new regional capitalist power. This pattern persists for other tropical commodities as well.

The tropical commodities became important items for mass consumption in the European core at a time when they were produced primarily under the direct control of colonial administrations. Thus they were initially incorporated into intranational trade within colonial blocs. As the system of nation-states was completed with the former colonies gaining their independence, and as the model of national regulation was extended within this system, tropical commodities increasingly entered into international trade. The geographic pattern of trade remained the same; only the form of its political regulation changed. The tropical commodities, which had been produced in the colonies for metropolitan consumption, became by default the major exportable products of the newly emerging nation-states, and export taxes on them became a major source of government revenues. At the same time, the tropical nature of these commodities gave their producers a potentially high degree of market power. The colonial legacy of these new nation-states was symbolized by their dependence on tropical community exports traded with former colonial powers, which earned low and decreasing prices on world markets. They reacted by organizing collectively, in a logical extension of their individual anticolonial movements, to exert their market power, restructure world markets, and improve their terms of trade.

THE REGULATION OF THE WORLD COFFEE MARKET DURING THE SECOND FOOD REGIME

After World War II, with the European market reopened, demand exceeded production, and prices rose rapidly to a peak in 1954. Brazil expanded production south into Paraná, where it was more susceptible to frosts, a move that had enormous repercussions for the postwar coffee market. The European countries expanded production in their African colonies, and independent African producers also expanded their output. All of this led to an oversupply of coffee after 1954 and steadily falling prices (Pan American Coffee Bureau 1966). Under these conditions, Brazil and Colombia, the two largest exporters, began efforts to organize a producers' cartel. In 1959 a producers' International Coffee Agreement (ICA) was signed. It was extended in 1960, by which time it included producers (or their colonial representatives) accounting for 94 percent of world production.[6] Still, this ICA had two major problems: Export quotas were set too high to maintain the world price, and some producers, having agreed to the quotas, proceeded to overship them. Thus, the agreement slowed the price decline but did not halt it (Finlayson and Zacher 1988, 153; Bilder 1963, 338).

The United States, as the hegemonic power in this period, was a strong supporter of free trade, but these actions by producers, along with the Cuban Revolution, convinced the Kennedy administration that cooperation with coffee producers was in the U.S. interest. The administration was concerned that a continued decline in coffee prices would lead to economic instability and the spread of communism throughout Latin America. The decolonization movement in Africa raised the threat of anti-imperialistic regimes being established there as well. After President Kennedy announced that the United States was willing to cooperate with commodity-producing countries to stabilize world market prices in order to promote the economic development of Latin America, as part of the Alliance for Progress, a new International Coffee Agreement, including both producing and consuming countries, was signed in August 1962.[7] This ICA set export quotas intended to maintain coffee prices at their late-1950s levels and obligated the consuming countries to enforce it by monitoring their imports to ensure that exporters adhered to the quotas (Fisher 1972, chap. 2).

The TNCs that dominated the U.S. market were generally supportive of the ICA, even though it meant somewhat higher producer prices. Their interests were represented by the National Coffee Association, dominated by the coffee TNCs and large commodity traders who dealt in coffee, which was instrumental in persuading the U.S. Congress to pass the legislation needed to implement the ICA. The NCA members shared the ideological positions of the Kennedy administration, and they were concerned with any disruptions of their supplies that might result from Third World revolutions, and thus favored U.S. participation to defend their interests against unilateral actions by producers (Fisher 1972, chap. 2; Payer 1975).[8]

The ICA was renewed with some modifications in 1968, but negotiations to renew it again broke down in 1972. The period of the second food regime was thus characterized by increasingly effective collective action by producers, culminating in a ten-year period with a politically regulated world market. It is generally agreed that regulation of the market by the ICA between 1962 and 1972 led to higher prices for coffee-producing countries than they would have obtained under a free market (Geer 1971, 171–88; Galloway 1973). The ICA therefore came to be regarded by both consuming and producing states as a form of "disguised aid," a means of transferring resources to Third World countries that did not require direct legislative approval.

Study of the world coffee market during this period reveals the two contradictory tendencies identified by Friedmann and McMichael with the second food regime: the spread of the model of national regulation of economic production and the increasing globalization of production organized by the TNCs. But because national regulation of the production of such tropical commodities as coffee led logically to attempts at international regulation of their trade, and because Third World state-capitalist alliances arose to challenge the TNCs, these dynamics played out differently in the world coffee market than in the food regime.

I have demonstrated above the ways in which national regulation led to international regulation. As Third World nation-states embarked on state-led industrialization and development strategies, the first obstacle they encountered was heavy dependence on tropical commodities and deteriorating terms of trade.

The need for increased foreign exchange earnings made commodity export policies an essential part of development strategy, and the concentration of market shares and the colonial legacy provided the economic and political bases for collective action. As the world system of nation-states was completed during this period, international organizations were created to administer a variety of new international regimes. The Third World attempted to use the rules of these new regimes to improve their economic positions in the world economy; this is illustrated specifically by the negotiation of the ICAs and more generally by the efforts of the United Nations Conference on Trade and Development (UNCTAD) (Krasner 1985). In the ICAs, Third World states succeeded in replacing the private regulation of the world market organized primarily by TNCs with a public regulatory regime, the rules of which were enforced by a multinational institution, the International Coffee Organization (ICO).

Contradictions between national and international regulation also emerged under this new regime. The most important multilateral decisions made by the ICO were on export quotas, which became divisive issues as each producer argued for a larger quota at the expense of the others. This was exacerbated within the ICA by the selectivity issue; the quota system limited not only the exports of each producer, but also the availability of each of the different types of coffee. When demands for different types were out of balance with their availability under the quotas, further frictions arose. Thus, no producing country was entirely happy with its quota under the ICA, and producers tended to split into blocs—Brazil and Colombia versus the Central American milds producers versus the African robusta producers (Khan 1978; Fisher 1972, chap. 7).

The increasing control of the TNCs over the final stages of coffee production in this period was manifested in the consolidation of their control over core consumer markets, where they significantly increased their market shares in roasted and instant coffee at the expense of smaller regional roasters. In the U.S. market, the four-firm concentration ratio for sales of roasted coffee increased from 46 percent in 1958 to 69 percent in 1978, while the number of independent roasters decreased from 261 in 1963 to 162 in 1972 (UNCTAD 1984, Table 11; UNCTC 1981, 84). Similar increases in concentration occurred in the major European markets (UNCTAD 1984). This trend was more dramatic for instant coffee, which involves a more highly industrialized final processing component. This final processing makes the retail price of a pound of green coffee processed into instant about 10 percent higher than a similar pound sold roasted and ground (UNCTAD 1984, Table 22). Because instant coffee is more capital intensive, its production is also more highly concentrated. The four-firm concentration ratio for instant coffee sales in the U.S. market increased from 72 percent in 1958 to 91 percent in 1978, and similar increases also occurred in the European markets (USFTC, 1966, Table 3; UNCTAD 1984, Table 13).

The increasing globalization of production was manifested in the TNCs' moves into Third World markets. The most significant move was made by Nestlé, which had pioneered the technology for spray-dried instant coffee in the

1930s. After the war, they began to establish subsidiaries to produce instant coffee in the Third World, first in Latin America, and then in Africa. By 1976, Nestlé had subsidiaries in twenty-one Third World countries (UNCTC 1981, 84). The first African subsidiary was established in Côte d'Ivoire to produce for markets in Greece and the Middle East. Location in Côte d'Ivoire gave Nestlé access to broken beans which could not be exported, at a discount price, and a waiver of export taxes for the first twelve years of operation. This location also gave Nestlé a foothold in the local market, where very little coffee was consumed when the plant first opened. But through promotional campaigns and discount offers, similar to those used in the oligopolistic competition among TNCs in the core markets, they were able to expand greatly the consumption of instant coffee in Côte d'Ivoire, as well as in other large markets in the region, such as Nigeria (Masini et al. 1979, 113–30; Dinham and Hines 1984, 61–67).

In response to these trends, the major producing countries attempted to establish their own instant coffee industries. Brazil took this a step farther and attempted to compete on the U.S. market.[9] The state created incentives to encourage instant coffee production by local capitalists, including access to broken beans and non-export-quality coffee, and state purchase guarantees. Then it exempted instant coffee producers from the export taxes that all green coffee exporters had to pay. This enabled Brazilian instant coffee firms to land instant coffee powder in the United States at a price significantly lower than the price at which it could be produced by factories in the United States buying green Brazilian coffee. In addition, this instant, which was produced entirely from arabica coffee, had a taste advantage in the U.S. market over the arabica-robusta blends used by most U.S. producers. Brazilian producers increased their share of the U.S. market from about 1 percent in 1965 to 14 percent in 1967.

This policy created a division among the coffee TNCs. The large coffee traders, who handled much of the green coffee ultimately processed by the TNCs, strongly objected because the Brazilian instant bypassed them and could be sold directly to TNCs and other roasters. The large diversified roasters, led by General Foods, were also opposed because the availability of the cheap Brazilian instant lowered the barriers to entry into the instant coffee market for other diversified food companies. Other large coffee firms, led by Hills Bros., along with most smaller regional roasters, supported the Brazilian policy because it gave them a cost advantage in competing with the larger TNCs. Hills Bros., at the time the fourth-largest coffee company in the United States, shut down its instant coffee operation in the United States and relied exclusively on buying and repackaging the Brazilian product. The U.S. government, and specifically the State Department, which was responsible for the ICA negotiations, sided with General Foods and the other TNCs which had helped it sell the first ICA to Congress. This controversy coincided with the renegotiation of the ICA in 1968, and the United States made its agreement to the ICA contingent on resolution of this "unfair trading practice" by Brazil. The United States, which at this same time was engaged in a massive subsidy operation to increase its wheat exports under the guise of food aid, succeeded in imposing a regulation prohibiting Third World producers from doing exactly the same thing to increase their industrialized product exports. However, the hypocrisy of this position was apparently recognized only in the Third World.

Thus, the breakdown of negotiations for a further renewal of the ICA at the end of the second food regime was due to factors specific to coffee as well as to the larger crisis of the food regime. First, the instant coffee problem still had not been fully resolved. The Brazilian government had pledged to end its "unfair" practices, but its compliance was inconsistent (Marshall 1971, 1972). Second, frosts in Paraná in 1969 and 1972 had reduced Brazil's exports and driven up the world market price. The smaller producers, particularly the African countries, had the capacity to export more coffee at a higher price under these conditions than they could hope to obtain quotas for under a new ICA, and they had no interest in agreeing to a new quota system. Third, the United States refused to acknowledge the effects of the breakdown of the Bretton Woods monetary system on Third World producers. World coffee prices, as well as the prices of most other primary commodities, were denominated in U.S. dollars, and the subsequent devaluation of the dollar meant a decrease in price for coffee exported to other core markets. Producers demanded an upward revision of the target price range that quotas were meant to defend, to compensate for this loss, but the United States refused to agree to this (Payer 1975; Brown 1980, 33–34; Marshall 1972).

Thus the breakdown of the ICA, as well as the larger crisis in the food regime, was partially caused by the globalization of capital organized by the TNCs, which overwhelmed national economic regulation and the incipient forms of international regulation. But it was also precipitated by the organized opposition of Third World commodity-producing states, which attempted to form alliances with local capitalists to challenge the TNC strategy. The 1973–1974 "energy shock," a key part of the crisis in the postwar food regime, was itself a product of collective action by commodity producers in OPEC. OPEC provided one model for collective action by Third World producers in the post-1973 struggle over the global reorganization of economic activity. The collective action by coffee producers, and by producers of other primary commodities, provided another, which led to UNCTAD's efforts to establish the Integrated Program for Commodities (IPC).

Commodity agreements with the participation of both producers and consumers were signed for tin and sugar in 1953. The ICA was signed in 1962, and an International Cocoa Agreement was concluded in 1972. Producers' agreements were negotiated for jute in 1965, hard fibers in 1967, and tea in 1969, all under the auspices of the Food and Agriculture Organization; but these agreements were much less effective in regulating these markets than the commodity agreements that included consumers. An organization of banana exporters was formed in 1974, but it had difficulty in getting the cooperation of all the major exporters (ironically, one of these was Colombia, one of the strongest supporters of the

ICA). Bauxite and copper producers also formed organizations in the mid-1970s (Brown 1980; Finlayson and Zacher 1988). This activity was driven by a similar set of circumstances: Third World producers were heavily dependent on primary commodity exports, often sent mainly to a former colonial power, for which they received low and unstable prices on world markets often dominated by TNCs. This movement culminated in the Sixth Special Session of the UN General Assembly in April and May 1974, which passed the "Declaration on the Establishment of a New International Economic Order." UNCTAD subsequently identified ten core commodities (cocoa, coffee, copper, cotton, jute, rubber, sisal, sugar, tea, and tin) for which it proposed a pooled international buffer stock arrangement combined with a compensatory financing mechanism, to stabilize the export earnings of developing countries. This proposal was the centerpiece of the IPC. Not surprisingly, the core countries, led by the United States, argued that this program went too far and would be far too costly, and refused to support it (Brown 1980). This confrontation signaled the beginning of a new period of struggle over forms of regulation of the world economy, which would be carried out under different conditions.

THE REGULATION OF THE WORLD COFFEE MARKET IN A FREE MARKET WORLD

The struggle over the shape of a new food regime and a new world economic order involves, as pointed out by Friedmann and McMichael, increasingly globalized agro-industrial production organized by TNCs and an insistence on free trade principles by core states. This has led, on one hand, to increased resistance by core states and TNCs to cooperate with collective efforts by Third World producers to regulate tropical commodity markets; and, on the other hand, to increased competition among Third World producers to expand and diversity their exports.

In the period following the end of the second food regime, the dependence of many poor Third World producers on tropical commodity exports has not significantly declined, but their market power has. Data for coffee are shown in Table 8.4, which shows the export dependence on coffee and market shares of total world exports for all exporters with at least a 0.1 percent market share in either 1967 or 1988–1989. In 1967 six countries depended on coffee for more than 50 percent of their total export value, and they accounted for 28.1 percent of total coffee exports; twelve countries depended on coffee for more than 30 percent of their exports, and they controlled 77.8 percent of coffee exports. By 1988–1989, five countries were still dependent on coffee exports for more than 50 percent of their export value, but they accounted for only 11.2 percent of coffee exports; ten countries still depended on coffee for more than 30 percent of their exports, but they controlled only 35.5 percent of the market.

Further, there were almost as many countries where export dependence had significantly increased as there were countries where it had significantly de-

Table 8.4
Dependence of Major Coffee Producers on Coffee Exports, 1967 and 1988–1989

Country	Coffee as % of Country's Exports		Market Share as % of Total Coffee Exports		GDP Per Capita	
	1967	1988-89	1967	1988-89	1967	1989
Uganda	53.5	96.1	4.4	2.9	87	260
Burundi	84.8	82.9	0.6	0.9	45	206
Rwanda	55.3	80.1	0.4	0.9	42	310
Ethiopia	55.6	61.5	2.5	2.9	61	126
El Salvador	47.7	59.6	4.5	3.6	261	1250
Tanzania	15.4	35.7	1.5	1.1	66	115
Guatemala	35.2	32.4	3.2	3.6	285	938
Madagascar	31.6	31.9	1.5	1.0	103	219
Colombia	63.2	30.5	14.6	17.8	313	1227
Nicaragua	14.6	30.4	1.0	0.8	339	258
Central African Republic	20.0	27.9	0.3	0.4	127	365
Kenya	29.5	23.8	2.0	2.6	111	356
Honduras	11.4	21.2	0.8	2.1	226	984
Costa Rica	38.1	20.7	2.5	2.9	394	1780
Cote d'Ivoire	32.5	18.8	4.8	5.6	223	804
Zaire	6.3	15.2	1.2	2.0	178	255
Cameroon	28.0	14.8	1.8	1.8	126	955
Papua New Guinea	19.5	10.3	0.5	1.6	n.a	936
Togo	10.6	9.1	0.2	0.2	98	391
Dominican Republic	10.9	8.5	0.8	0.8	251	953

Table 8.4 Continued

Haiti	na	8.2	na	0.2	79	368
Ecuador	20.2	7.3	1.8	1.8	218	979
Brazil	44.3	5.9	33.2	21.9	273	3035
Sierra Leone	0.7	5.5	–	0.1	142	232
Panama	1.6	5.4	0.1	0.2	553	1933
Peru	3.9	4.3	1.4	1.4	263	1982
Malawi	–	4.2	–	0.1	50	192
Guinea	na	3.8	na	0.2	96	437
Paraguay	–	2.7	–	0.2	211	1052
Indonesia	6.6	2.5	2.0	5.6	94	523
Mexico	5.4	2.4	2.5	5.8	520	2314
Zimbabwe	na	2.1	na	0.3	222	624
Bolivia	–	2.1	–	0.2	176	608
India	1.6	1.4	1.2	2.0	81	326
Angola	51.9	1.2	5.6	0.3	68	716
Jamaica	0.3	1.2	–	0.1	493	1601
Cuba	–	0.9	–	0.6	n.a	1562
Liberia	1.6	0.8	0.1	–	272	497
Philippines	–	0.7	–	0.6	259	729
Thailand	–	0.3	–	0.7	140	1259

Source: UNCTAD, *Handbook of International Trade and Development Statistics*, (New York: UN, various years); the 1967 data on coffee exports are from the 1976 Handbook, Tables 4.3(D) and 4.4; the 1967 GDP per capita data are from the 1969 Handbook, Table 6.1A; the 1988-89 data are from the 1991 Handbook, Tables 4.3 and 4.4 for coffee exports, and Table 6.1 for GDP per capita. All coffee export data are for SITC 071, which includes green coffee, coffee extracts and instant coffee. Market shares shown are as percentages of total developing country coffee exports (developing countries account for 96% of coffee exports in 1967, and 88% in 1988-89). A – means less than 0.1%.

clined. Seven countries with more than 20 percent dependence on coffee exports in 1967 registered declines of more than 10 percent in their export dependence between 1967 and 1988–1989. Brazil is the only country in this group for which the decline was accounted for by an increase in manufactured exports. Three others, Angola, Cameroon, and Ecuador, decreased their dependence on coffee by increasing their dependence on oil; for Colombia the decline in coffee dependence was accounted for by increases in oil and coal exports. Costa Rica decreased its dependence on coffee by expanding banana exports, and Côte d'Ivoire decreased its coffee dependence by expanding cocoa exports. Six countries with at least 10 percent dependence in 1967 significantly increased their dependence on coffee exports by 1988–1989, and their stories are equally dismal. Four of them, Uganda, Tanzania, Nicaragua, and El Salvador, had significant levels of dependence on cotton exports in 1967 and became more dependent on coffee because of the expansion of U.S. cotton exports, the fall in the world price for cotton, or disintegration of their cotton production systems. Rwanda became more heavily dependent on coffee after the tin price crashed in the mid-1980s. And Honduras became more dependent on coffee by expanding its coffee production. Six additional countries remained about equally dependent on coffee in 1988–1989 as they had been in 1967, at levels over 20 percent (Burundi, Ethiopia, Guatemala, Madagascar, Central African Republic, and Kenya).

Market shares in the world coffee market have been dispersed by two tendencies. First, World Bank/IMF policies of export promotion, which intensified in the early 1980s in response to the debt crisis, drove coffee producers to diversify and expand their exports. But as the example of coffee shows, this often did not decrease their dependence on primary commodities; it just shifted their dependence from one primary commodity to another, or spread it over a few more. Second, the relatively high value of coffee on the world market, ironically maintained by the collective efforts of the producing countries, encouraged new producers to enter the market. This tendency is exemplified by Honduras, until recently a minor Central American producer, whose market share is now approaching that of Costa Rica; and by Thailand, which became a coffee exporter during this period as part of a larger strategy of export diversification, through which it has emerged as one of the leading NACs. A combination of these two tendencies is exemplified by Mexico and Indonesia, which have increased their market shares while decreasing their dependence on coffee exports. Both have expanded their coffee production and increased their exports of manufactured goods, while Mexico has also become a major oil exporter. These two countries also deserve inclusion in the class of NACs, to the extent that they have diversified into new and valuable agricultural exports.

The tendencies toward globalization of agricultural production, as identified by Friedmann and McMichael, created pressures on Third World agricultural producers to expand and diversify their exports. A few NICs (e.g., Brazil and Mexico) have escaped heavy dependence on agricultural or primary commodity exports, while a few NACs (e.g., Thailand, Mexico again, and possibly Indonesia) have succeeded in finding profitable niches in the new global agricultural production order. But these four countries are really the only ones, out of the forty countries listed in Table 8.4, that can be identified as success stories. In 1967 there was a basis for strong collective action by a relatively small group of coffee exporters who controlled a large share of the market, but by 1990 the dispersion of market shares had begun to undermine this possibility. The states with the greatest market power and capacity to intervene in the market now tend to have the least interest in making intervention in the world coffee market a major focus of their economic policies. Despite this tendency, coffee-producing states continued to intervene collectively in the world coffee market through the 1980s.

The breakdown of the ICA in 1972, and the failure of the Brazilian instant coffee strategy, led coffee producers to adopt more aggressive forms of market intervention in the post-ICA world. This intervention began in April 1972 with the formation of the Geneva Group. By that time it had become clear to the producers that the ICA negotiations were not going to succeed, and fourteen producers, led by Brazil, Colombia, Côte d'Ivoire and Portugal, met to draft a new producers' agreement. These countries pledged to maintain exports at the 1972 quota levels, in order to hold prices at the relatively high levels they were then receiving. As a complement to this strategy, Brazil, Colombia, and Côte d'Ivoire formed a corporation, called Café Mundial, in 1973 to operate a multinational buffer stock for coffee. Since the Geneva Group and Café Mundial were private agreements between producing states, not much is publicly known about the details and the scale of their operations, but various shifts in world market prices during 1973–1974 were attributed to them. The coincidence of these efforts with UNCTAD's IPC proposals worried the core states; it seemed as if Third World states, anticipating a core refusal to go along with the IPC, were going to attempt to implement it unilaterally. At this same time, a very severe frost hit Paraná in June 1975, destroying over half of Brazil's coffee crop; and in August 1975, civil war broke out in Angola, then one of the largest producers, disrupting its coffee exports. This led the TNCs to have renewed worries about the stability of their supplies (Finlayson and Zacher 1988, 158–59; Economist Intelligence Unit 1987, 18; Wasserman 1972).

Under these conditions, the United States and other core consuming states decided that a renewed ICA might be in their best interests. But under these conditions, it is harder to explain why the producing states would have had an interest in a new ICA. It seems likely that most of them realized that the shortage of supplies would only be temporary, and that quotas would probably be necessary to stem the fall in price which would happen as Brazilian production began to recover. At the same time, it was easier for them to agree on quotas which would be immediately suspended because the world price was at such a high level, and would thus have no immediate impact on their exports. The agreement went into effect in 1976, with the quotas suspended.

Prices dropped fairly rapidly after 1976–1977, as Brazilian exports recovered and other countries reduced their stocks. Producers began to call for a reimposition of the quotas to stop this decline, but consuming states preferred to see prices fall to their 1960s' levels before agreeing to enforce the quotas.[10] The failure to reimpose quotas led producers to undertake further collective action to affect world prices. In 1978 Brazil, Colombia, Venezuela, Mexico, El Salvador, and Honduras formed the Bogotá Group.[11] This time, the producing states attempted to intervene directly in the major coffee markets of the consuming countries, in New York and London. New York is the major port of entry for U.S. coffee imports, headquarters of most of the large commodity transfers, and the location of a large spot market, where lots of coffee held by the traders are available for immediate purchase. It is also the location of the world's largest futures market for arabica coffees: the New York Coffee, Sugar, and Cocoa Exchange. London is the location of the London Terminal Market, the largest futures market for robusta coffees.

Once again, the details of the Bogotá Group's operations remain largely a matter for speculation in the coffee trade. Apparently, however, they bought coffee in New York and held it off the market to drive up spot prices, on which prices of many other coffee purchase contracts were based. Additionally, they apparently bought coffee futures on the New York and London markets to create the impression that coffee might be in short supply several months in the future. This was calculated to cause TNCs to purchase more coffee through commodity traders in order to ensure the stability of their supplies, further driving up world prices. In May 1980, the Bogotá Group went public, forming a corporation, called Pancafé, which applied for membership as a registered trader on the New York and London futures markets. But by August 1980, rumors were spreading that Pancafé was in financial trouble because it was holding coffee and futures contracts, and increased production stimulated by the 1976–1977 price boom was driving the world price down. In September, consuming states agreed to reinstitute quotas to stop the decline in prices in return for the liquidation of Pancafé. The 1976 ICA was set to expire in 1982, but it was extended for one year, and a new ICA was negotiated in 1983. Quotas were continuously in force under these ICAs through 1989, except for a brief period following another Brazilian frost in 1986, when they were suspended to control the resulting price rise. Negotiations to renew the ICA broke down in 1989, when the 1983 ICA expired, and neither producers nor consumers seem interested in negotiating a new one at present.

The reimposition of an internationally regulated market through the 1980s in instructive, both because of its similarities with the situation in 1962, which led to the first ICA, and because it occurred within a radically different international context. First, I will discuss the similarities. In 1962, as in 1980, geopolitical concerns were important in motivating participation in market regulation by the core states. In 1962 it had been the Cuban Revolution, the threat of spreading revolution in Latin America, and the decolonization movement in Africa, which caused core states, led by the United States, suddenly to become interested in promoting the economic development of the Third World. In 1980 it was the Sandinista revolution in Nicaragua, the civil wars in El Salvador, Guatemala, and Angola, and escalating guerrilla warfare in Colombia, which caused similar concerns. In this geopolitical context, a mechanism for the provision of disguised aid to specific Third World states by the United States and other core states became an important instrument of foreign policy. In 1980, as in 1962, core states' geopolitical concerns became linked to TNC concerns over the stability and prices of their coffee supplies. In both cases, the actual disruption of coffee supplies, because of political upheavals, was compounded by fears that future supplies might also be cut off by U.S. import bans, as had been applied against Cuba, or against Uganda under Idi Amin, in the late 1970s. The TNCs supported market regulation to stabilize prices and approved of the transfer of aid to the coffee-producing countries, if it was necessary to stabilize their political situations. Finally, in 1962 and in 1980, core states and TNCs alike were concerned about the possible effects of collective action by Third World producers, who were attempting to impose their own forms of regulation on the world market. Thus, a form of regulation over which the core states could exercise some degree of control seemed to them to be a preferable alternative.

But the 1962 and 1980 agreements to impose internationally controlled quota restrictions on the market are also remarkable because they occurred in different international contexts. In 1962, during the second food regime, the established and newly emerging Third World states were engaged in collective action to change the terms of trade, in what was still a largely colonial pattern of trade. This collective action resulted in a wide variety of international agreements to regulate trade, and the ICA was just one of these. In 1980, with a new food regime emerging, globally organized by the TNCs, the ability of Third World producers to regulate their own national agricultural production was significantly reduced. Yet Third World coffee producers had demonstrated a remarkable ability collectively to control their own exports and to coordinate market interventions that reached past the world market and into the markets of the core consuming countries. The main reason that regulation of the world coffee market in the 1960s took a different form than the regulation of the food regime, as it has been described by Friedmann and McMichael, was the fact that coffee was a tropical commodity produced and traded on the old colonial pattern. The main reasons that a similar form of regulation was maintained through the 1980s were that coffee, despite a degree of restructuring and globalization of its production, still exhibited the colonial trade pattern with green coffee production remaining under national control; and that coffee producers, with twenty years of experience, had developed the state capacity and the institutional framework that made continued collective action possible.

CONCLUSIONS

I have argued that the tropical commodities, and the poor Third World countries which remain heavily dependent on producing and exporting them, have been overlooked in analyses of the organization of the world-economy. As a result, these analyses fail to identify one of the main dynamics shaping the world agricultural system—the legacy of colonialism, manifested in the neocolonial production and trade patterns of these commodities. This colonial legacy, and the nature of tropical commodity chains, create the political and economic bases for collective resistance by Third World states to the global restructuring projects of the TNCs and their core states. Collective actions by Third World states have succeeded to some degree in establishing international institutions that can impose political regulations on tropical commodity markets. And to the extent that they have succeeded in these efforts, Third World states have created the state capacities, institutional frameworks, and political expectations that will enable them to continue to influence the structure of the world economy.

I have made this argument by focusing primarily on coffee as an exemplary tropical commodity. How representative is coffee? I have argued that all tropical commodities share a basic set of characteristics, but they are different commodities with distinct commodity chains, exported and imported by different sets of countries. My argument is that, although the specifics of each commodity will be somewhat different, the underlying dynamic is basically similar. Tropical commodity producers are all pushed toward organizing collectively to confront the TNCs and to impose political regulations on world markets. Coffee illustrates these tendencies more clearly than most other tropical commodities because of its importance in world trade, and it is thus the best candidate for a case study to analyze this dynamic.

I have developed this analysis through comparisons with food regime theory, as elaborated by Friedmann and McMichael. I have argued that, although food regime theory provides many valuable insights into the organization of world agricultural systems, a consideration of the tropical commodities adds three specific insights regarding the limits of globalization. First, some agricultural commodities are more susceptible to the tendencies toward global sourcing and industrialized production than others. Tropical commodities continue to be produced and traded on a pattern established before the first food regime, and this pattern may persist alongside the consolidation of a third food regime. But globalization negatively affects the poor Third World countries which remain heavily dependent on tropical commodities to the extent that it creates oversupply and increased competition in the markets for these commodities. More generally, each restructuring of the world agricultural system since the colonial period has tended to create a few winners, which have been skillful or fortunate enough to find a profitable niche in the new order. The tendency for sociologists and others who study the Third World is to focus on these few cases and overlook the oppressed majority who are left farther behind with each restructuring (Wallerstein 1979).

Second, there has been a long history of collective struggles by Third World states, allied with Third World capitalists, to change the political structures of the world markets for these tropical commodities. Many of these political efforts were defeated by a combination of TNC efforts to escape regulation by globalizing their production processes and core state (primarily U.S.) efforts to impose a new regime of free trade on world markets. But the persistence of colonial patterns of production and trade, and the past history of Third World collective action, provide the economic and political bases for continued opposition to globalization. The international institutions created by the previous struggles (e.g., UNCTAD or the ICO) remain in place and provide an institutional base for cooperation by Third World producers. And the past successes create expectations for future organized opposition. The contradiction in the emerging food regime is thus not only that Third World states lose legitimacy if they are unable to provide food security, as McMichael (1992) argues, but also that they lose legitimacy if they are unable to impose some degree of regulation on world tropical commodity markets, in order to benefit the local capitalists who control their production and export.

Third, there are also tendencies at work in the global restructuring of agricultural production that undermine the further possibilities of collective action. The Uruguay Round of GATT and World Bank/IMF policies to promote free trade and open markets tend to reduce the concentration of market shares, encourage new producers to enter the market, and force tropical commodity producers to compete against one another. The emergence of NICs and NACs divided Third World producers by giving these states a stake in the newly emerging food regime that is not shared by the poorer states still dependent on tropical commodity exports. And the development of biotechnology, particularly the increase of appropriationism, opens new possibilities for TNCs to penetrate earlier stages of tropical commodity chains which have so far been resistant to industrialization. It was recently reported that a genetically engineered coffee tree, resistant to several common coffee diseases, has been developed. There has also been talk for several years of the possibility of developing a coffee tree that would produce decaffeinated beans. These developments might revolutionize the earlier stages of coffee production. But given the basic climatic requirements of coffee growing, there may be limits as to where these proprietary coffee trees could be grown. And given the propensity and capacity of the coffee-producing states to exercise control over their coffee production, it is not clear that even this type of appropriationism would put green coffee production under the control of the TNCs.

NOTES

Revised version of a paper presented at the Seventeenth Annual Conference of the Political Economy of the World System section of the American Sociological Association, Cornell University, April 15–17, 1993. I would like to thank Philip McMichael and Laura Enriquez for their very helpful comments on the earlier version of this chapter.

1. I have disaggregated UNCTAD's category, which includes oils and oilseeds with different trading patterns, by separating out soybeans and soybean oil, which are a key part of the livestock/food complex, and palm oil, which has been substituted for by temperate oils in the durable food complex. The main constituents of my new category of tropical oils and oilseeds are groundnuts and groundnut oil, copra and coconut oil, and palm kernels and palm kernel oil. These oils and oilseeds have also been subjected to substitutionism in the production of durable foods, but they are still exported primarily to the core (cf. Friedmann 1991a, 77).

2. The exceptions are natural rubber, which was one of the earliest tropical products to be partially replaced by a core-produced synthetic substitute, and sugar and oils, which have been substituted for in the durable food complex.

3. Note that this industrial production of instant coffee only occurs at the final stage of processing of already roasted coffee, and that the earlier processing stages of the commodity chain are not capital or technology intensive.

4. Coffee was fourth overall among primary commodities in the instability of its world market price, 1955–1981. Sugar and cocoa prices were even more unstable during this period, ranking first and second, respectively. All tropical commodities listed above experienced at least 7 percent average annual price fluctuations over the 1955–1981 period; for coffee, the average annual price fluctuation was 17.8 percent (Finlayson and Zacher 1988, Table 1.4, p. 7).

5. The United States is, of course, strictly speaking, not the former colonial power here; however, since Latin American independence, the United States has regarded Latin America as its "back yard" and has been its main trading partner.

6. Brazil, Colombia, Costa Rica, Cuba, Dominican Republic, Ecuador, El Salvador, Guatemala, Haiti, Honduras, Mexico, Nicaragua, Panama, Peru, and Venezuela were the signatories to the 1959 ICA; they were joined in 1960 by the newly independent "franc zone" countries of Cameroon, Central African Republic, Congo (Brazzaville), Côte d'Ivoire, Dahomey, Gabon, Malagasy Republic, and Togo, along with Portugal (for Angola) and the United Kingdom (for Kenya, Tanganyika, and Uganda) (Fisher 1972, 19–27).

7. Members of the 1962 ICA were, as exporting members: Brazil, Burundi, Cameroon, Central African Republic, Colombia, Congo-Brazzaville, Congo-Leopoldville, Costa Rica, Côte d'Ivoire, Cuba, Dahomey, Dominican Republic, Ecuador, El Salvador, Ethiopia, Gabon, Guatemala, Haiti, Honduras, India, Indonesia, Kenya, Malagasy Republic, Mexico, Nicaragua, Nigeria, Panama, Peru, Portugal (Angola), Rwanda, Sierra Leone, Tanganyika, Togo, Trinidad and Tobago, Uganda, Venezuela, and Yemen; and as importing members: Argentina, Australia, Austria, Belgium, Canada, Cyprus, Czechoslovakia, Denmark, West Germany, Finland, France, Israel, Italy, Japan, Luxembourg, Netherlands, New Zealand, Norway, Spain, Sweden, Switzerland, Tunisia, United Kingdom, United States, and the Soviet Union. Members accounted for 98 percent of world exports and over 94 percent of world imports (Fisher 1972).

8. Krasner (1973a, 1973b) argues that the TNCs were hurt economically by the first ICA, and thus, while they made supportive public comments about the agreement, they went along with it only at the urging of the State Department out of a sense of public duty. This argument seems implausible, given the difficulty that the U.S. administration had in getting the implementing legislation through Congress—this did not occur until 1965. John McKiernan, president of the National Coffee Association (NCA), appeared

at all of the Congressional hearings on various drafts of this legislation during the 1962–1965 period to voice the strong support of NCA members for the ICA. It seems unlikely that the legislation would have passed at all without this consistent support.

9. This account of the Brazilian instant coffee controversy is drawn from Fisher 1972, chap. 9; Cordell 1969; Krasner 1973a; Marshall 1969; and Lucier 1988, 139–47.

10. Green coffee prices in the New York market during the 1960s were in the range of from 40 to 50 cents per pound, and the 1976 ICA provided for an automatic reintroduction of quotas if the price reached 78 cents per pound. Prices peaked in early 1977 at over $3.00 per pound, and the producers sought to hold them at their mid-1978 level of about $1.50.

11. The details presented here on the operations of the Bogotá Group and Pancafé are drawn primarily from Economist Intelligence Unit 1987, 21–27; and from Marshall 1980.

REFERENCES

Bilder, Richard B. 1963. "The International Coffee Agreement: A Case History in Negotiation." *Law and Contemporary Problems* 28, 2: 328–91.

Brown, Christopher. 1980. *The Political and Social Economy of Commodity Control.* New York: Praeger.

Cordell, Arthur J. 1969. "The Brazilian Soluble Coffee Problem: A Review." *Quarterly Review of Economics and Business* 9: 29–38.

Dinham, Barbara, and Colin Hines. 1984. *Agribusiness in Africa.* Trenton, N.J.: Africa World Press.

Economist Intelligence Unit. 1987. *Coffee to 1991: Controlling a Surplus.* London: Economist.

Finlayson, Jock, and Mark Zacher. 1988. *Managing International Markets: The Developing Countries and the Commodity Trade Regime.* New York: Columbia University Press.

Fisher, Bart. 1972. *The International Coffee Agreements: A Study in Coffee Diplomacy.* New York: Praeger.

Friedmann, Harriet. 1978. "World Market, State, and Family Farm: The Social Bases of Household Production in the Era of Wage Labor." *Comparative Studies in Society and History* 20, 4: 545–80.

———. 1982. "The Political Economy of Food: The Rise and Fall of the Postwar International Food Order." In *Marxist Studies*, edited by M. Burawoy and T. Skocpol. *American Journal of Sociology* 88 (Supplement): 248–86.

———. 1991a. "Changes in the International Division of Labor: Agri-Food Complexes and Export Agriculture." In *Towards a New Political Economy of Agriculture*, edited by W. Friedland et al., 65–93. Boulder, Colo.: Westview Press.

———. 1991b. "New Wines, New Bottles: The Regulation of Capital on a World Scale." *Studies in Political Economy* 36: 9–42.

Friedmann, Harriet, and Philip D. McMichael. 1989. "Agriculture and the State System: The Rise and Decline of National Agricultures, 1870 to the Present." *Sociologia Ruralis* 29, 2: 93–117.

Galloway, L. Thomas. 1973. "The International Coffee Agreement." *Journal of World Trade Law* 7, 3: 354–74.

Geer, Thomas. 1971. *An Oligopoly: The World Coffee Economy and Stabilization Schemes.* New York: Dunellen.

Gereffi, Gary, and Miguel Korzeniewicz. 1990. "Commodity Chains and Footwear Exports in the Semiperiphery." In *Semiperipheral States in the World Economy*, edited by William Martin, 45–68. Westport, Conn.: Greenwood Press.

Goodman, David. 1991. "Some Recent Tendencies in the Industrial Reorganization of the Agri-Food System." In *Towards a New Political Economy of Agriculture*, edited by W. Friedland et al., 37–64. Boulder, Colo.: Westview Press.

Khan, Kabir-ur-Rahman. 1978. "International Coffee Agreement 1976: Issues of Selectivity, Regulation, and Reciprocity." *Food Policy* 3, 3: 180–90.

Krasner, Stephen. 1973a. "Business-Government Relations: The Case of the International Coffee Agreement." *International Organization* 27, 4: 495–516.

———. 1973b. "Manipulating International Commodity Markets: Brazilian Coffee Policy 1906–1962." *Public Policy* 21, 4: 493–523.

———. 1985. *Structural Conflict: The Third World against Global Liberalism*. Berkeley: University of California Press.

Lucier, Richard. 1988. *The International Political Economy of Coffee*. New York: Praeger.

McMichael, Philip D. 1992. "Tensions between National and International Control of the World Food Order: Contours of a New Food Regime." *Sociological Perspectives* 35, 2: 343–65.

McMichael, Philip D., and David Myhre. 1991. "Global Regulation vs. the Nation-State: Agro-Food Systems and the New Politics of Capital." *Capital & Class* 43: 83–105.

Marshall, C. F. 1969. "World Coffee Problems." *Bank of London and South America Review* 34: 616–23.

———. 1971. "Coffee: The State of the Market." *Bank of London and South America Review* 50: 64–74.

———. 1972. "Coffee: The Market in 1971 and the Present Position." *Bank of London and South America Review* 63: 127–34.

———. 1980. "Coffee in 1979: The Effect of the Bogotá Group." *Bank of London and South America Review* 14, 1: 2–6.

Masini, Jean, Moises Ikonicoff, Claudio Jedlicki, and Mario Lanzarotti. 1979. *Multinationals and Development in Black Africa: A Case Study in the Ivory Coast*. Westmead, Farnborough, Hampshire, England: Saxon House, Teakfield Ltd., and the European Center for Study and Information on Multinational Corporations.

Pan American Coffee Bureau. 1966. *Annual Coffee Statistics 1966: Thirtieth Anniversary Edition*. New York: PACB.

Payer, Cheryl. 1975. "Coffee." In *Commodity Trade of the Third World*, edited by C. Payer, 154–68. London: MacMillan.

UNCTAD (United Nations Conference on Trade and Development). 1984. *The Processing and Marketing of Coffee: Areas for International Cooperation*. New York: United Nations.

UNCTC (Centre on Transnational Corporations). 1981. *Transnational Corporations in Food and Beverage Processing*. New York: United Nations.

U.S. Federal Trade Commission (USFTC). 1966. *Cents-Off Promotions in the Coffee Industry*. Washington, D.C.: Bureau of Economics Staff Report.

Wallerstein, Immanuel. 1979. "Dependence in an Interdependent World: The Limited Possibilities of Transformation within the Capitalist World-Economy." In *The*

Capitalist World-Economy, edited by Immanuel Wallerstein. Cambridge, England: Cambridge University Press.

Wasserman, Ursula. 1972. ''1972 Geneva Coffee Agreement.'' *Journal of World Trade Law* 6, 4: 612–14.

9

On Global Pond: International Development and Commodity Chains in the Shrimp Industry

Mike Skladany and Craig K. Harris

This chapter examines the emergence of penaeid shrimp (*Penaeus* sp.) both as a vehicle of international development and as a commodity with a particular system of production and exchange. Another paper (Harris and Skladany 1993) provided an overview of earlier capture fisheries and historic pond culture activities and framed the recent restructuring of the shrimp industry toward intensive pond culture. Here, we examine the explosive emergence of tropical shrimp pond culture and its position in the global commodity system of production and exchange. The chapter then discusses key factors associated with the rapid growth of shrimp pond culture in Asia and Latin America. Finally, consideration is given to some of the consequences of, and the contradictions underlying, development of aquatic resources in this international context.

THE EMERGENCE OF A GLOBAL SHRIMP ECONOMY

The harvest, culture, and consumption of aquatic organisms have historically been matters of acute international attention, and the exploitation of capture fisheries stocks has always been difficult to rationalize. The resulting uncertainties inherent in capture fisheries have motivated many states, corporations, and entrepreneurs to intensify various forms of raising aquatic organisms in a more controlled environment. At present, pond-raised shrimp accounts for roughly one-third of the global production (Table 9.1). Technological advancements in breeding and nutrition (Pillay 1990, 425–44), favorable state policies (Sarig 1988, 333–64), investment and financial incentives, and increasing market demand for seafood in the United States, Japan, and Europe (Lee and Wickins 1992, 20–52) have provided an explosive set of conditions for the mercurial growth of pond raising of shrimp in Asia (e.g., Indonesia, Thailand, Vietnam,

Table 9.1
Global Shrimp Harvest, 1981–1991

Year	Harvest Capture	Culture	Total Harvest	Proportion Cultured
	1,000 Metric Tons*			Percent
1981	NA	NA	1,627	NA
1982	1,653	84	1,737	5
1983	1,684	143	1,827	8
1984	1,746	174	1,920	9
1985	1,925	213	2,138	10
1986	1,927	309	2,236	14
1987	1,821	551	2,372	23
1988	1,902	604	2,506	24
1989	1,869	611	2,480	25
1990	1,875	652	2,527	26
1991	NA	699	NA	NA

* - Live weight equivalent
NA - Not available
SOURCE: United States Department of Commerce (1992a).

and China) and in Latin and Central America (e.g., Ecuador, Mexico, Honduras, Brazil, and Colombia).[1]

The overall structure of development and investment was initially directed by transnational entities such as the World Bank, the Asian Development Bank, national banks, and bilateral aid agencies. These entities, as they did during the boom period of capture fisheries (e.g., Bailey 1988a, 1988b; Klausen 1968), acted as catalysts for private sector investment which is often "tied" between various intragovernmental actors (Meltzoff and LiPuma 1986; U.S. Department of Commerce 1992a–d). Over the past decade, foreign exchange requirements, favorable state policies and liberal investment incentives have attracted substantial investment by transnational and/or national firms and groups who are emerging as the dominant sets of actors in the production, financing and marketing of shrimp (U.S. Department of Commerce 1988, 1992a–d).

As investment has generally flowed from north to south, shrimp commodities generally flow from the south to the industrialized north. This pattern is also undergoing substantial restructuring in terms of capital flows, production modes, marketing, distribution, and consumption. The relative share of shrimp caught from the sea is gradually diminishing (see Table 9.1). Pond-raised shrimp find ready markets in a variety of diverse regions in the world including U.S.–based fast-food chains and as Asian imports into a variety of countries (Rosenberry 1990; *World Shrimp Farming*, March/April 1993, 13). In short, the earlier bilateral pattern of trade which was dominated by capture fisheries has become integrated along multilateral lines of regional production and exchange with the emergence of pond culture of shrimp. In pond culture, south-to-south, as well as south-to-north linkages are becoming more prevalent in the industry.

"THE BLUE REVOLUTION" IN SOUTH
AND SOUTHEAST ASIA

Parallel to the much-documented "green revolution" a quieter but corresponding "blue revolution" unfolded primarily in the maritime south and Southeast Asia beginning in the 1950s and accelerating throughout the 1960s (Bailey 1988a). The deployment of powerful production technologies was instigated by international aid agencies (e.g., World Bank, Asian Development Bank, German Technical Assistance Agency, and Norwegian Agency for International Development) within the social, political, and economic context of traditional fisheries in a series of Southeast Asian nations (Bailey 1988a; Goonatilake 1984; Maril 1983).

As a result of the opening up of new fisheries resources which were beyond the reach of traditional fisheries (Bailey 1988a; Panayoutou and Jetanavanich 1987), marine fisheries production rapidly grew in countries such as Thailand, India, and Indonesia only to level off in the 1980s (Food and Agriculture Organization 1989). With the general imposition of 200-mile exclusive economic zones (EEZs) in 1977 in anticipation of the ratification of the Convention on the Law of the Sea, regional industrial fishing powers with relatively small EEZs (such as Thailand) were forced to fish illegally in other countries and to intensify efforts within their own territorial waters (Panayoutou and Jetanavanich 1987). The intensity of efforts is such that a recent global assessment clearly shows that in *every* major fishing region of the world shrimp stocks are either over- or fully exploited (FAO 1989).

EARLY POND PRODUCTION EFFORTS

As a result, by the mid-1980s, capture shrimp efforts had leveled off in conjunction with some fluctuations in the harvest. Concurrent with the maritime expansion, although much more widely dispersed geographically, were efforts to raise shrimp in a controlled environment. Asian production of shrimp in ponds has been long standing. In what are known as *extensive* production systems, an enclosure is built close to the sea and tidal flows into and out of the enclosure provide stocking of shrimp, feed, and water exchange (Bailey 1992; Ling 1977; Pillay 1990). Generally, production in these systems is low. Up to 1980, culture of shrimp in extensive systems accounted for only 6 percent of world production (Table 9.2).

In the late 1970s, international agencies, such as the World Bank and the Asian Development Bank, began providing loans to governments for the development of shrimp pond production. At this time, the whole notion of "mariculture" was akin to a mythology of "farming the seas" for supplying humanity with much needed protein (Borgstrom 1967, 374–98). The following paragraphs review the early developments in Ecuador, Japan, Hawaii, and Taiwan.

After initial efforts in 1969, shrimp pond production in Ecuador increased to 5,000 hectares under production by 1979. At this time, Ecuador was beginning

Table 9.2
Shrimp Pond Production by Country, 1980–1990

Country	Year						
	1980	1985	1986	1987	1988	1989	1990
	1,000 Metric Tons#						
China	2	35	70	150	200	165	150*
Taiwan	5	33	65	75	35	20*	30
Indonesia	35	50	53	73	96	90	120*
Ecuador	9	27	36	38	70	45	73
Thailand	10	15	17	30	45	90	110*
Bangladesh	7	12	13	14	15	–	25
India	12	17	18	22	24	25	32
Philippines	1	28	30	35	36	50	30
Vietnam	4	7	7	7	25	30*	30
Japan	1	1	2	7	7	–	3
Peru	1	2	3	4	4	–	–
Panama	–	1	1	1	2	–	–
Others	7	16	18	21	25	49	29
Total	96.5	247.8	337.4	478.5	584.4	564.8	632.9
% of Global Production**	6%	13%	17%	26%	31%	31%	34%

- rounded off figures
* - estimated
** - estimates only
SOURCE: United States Department of Commerce 1992a.

to realize the potential for profit and export earnings from shrimp despite relatively extensive methods. Subsequently, the area devoted to shrimp culture rapidly increased from 5,000 hectares to 100,000 hectares by 1985. Of this 100,000 hectares, approximately 75,000 were coastal mangrove swamps that were destroyed in order to build ponds (Meltzoff and LiPuma 1986; U.S. Department of Commerce 1992d).[2]

In Ecuador's agricultural sector, a series of boom and bust cycles associated with bananas and cocoa provided the impetus for the private sector development of other export commodities. Close proximity to the major world market, the United States, and lack of competition from other countries and regions in the world ensured that Ecuadorean shrimp found high prices. Shrimp became the country's second leading export commodity next to oil, and by 1985 revenues from shrimp exports were estimated at over 400 million dollars per annum (Meltzoff and LiPuma 1986). While the coastal area suitable for shrimp cultivation is believed to be fully exploited at approximately 110,000 hectares, there is a growing tendency among the roughly 1,500 pond operators, groups, and companies to switch to more intensive production methods (U.S. Department of Commerce 1992b, 1992d).

Other early pond culture efforts began in Japan in the 1930s with the development of controlled breeding and pond culture techniques for *Penaeus japonicus*, the high-value kuruma prawn (Hanson et al. 1977, 7; Brown 1989, 381–86; Shigeno 1979). In Japan, however, constraints such as a lack of areas suitable for pond aquaculture and high labor costs acted against further development of the industry, and pond-raised shrimp accounts for only 7 percent of Japan's shrimp production (U.S. Department of Commerce 1992a).

Important to the development of the global pond shrimp production industry was Hawaii's position. In 1979 the Japanese firm IKKO leased land to develop pond production in kuruma prawns for export live to the Tokyo market. The farm eventually failed in 1983 despite prices as much as U.S. $40.00 per pound (U.S. Department of Commerce 1992a). Other early ventures in Hawaii included the Coca Cola–University of Arizona venture which failed in the mid-1980s due to a viral disease outbreak prior to their first harvest (U.S. Department of Commerce 1992a). Hawaii's Oceanic Institute began research into shrimp culture in 1978 focusing on various technical and economic aspects of shrimp farming and subsequently entered into a relationship with the University of Arizona to supply broodstock for distribution to farmers in the southwestern United States (U.S. Department of Commerce 1992a; World Shrimp Farming 1993, 2–10).

Efforts in Taiwan in the early 1970s were pioneering with respect to the eventual cultivation of shrimp in a controlled environment (Chen 1976, 111–22). Taiwanese aquacultural scientists successfully identified and developed methods for the mass production and culture of the black tiger shrimp (*Penaeus monodon*). Mass production of seed stock led to the development of hatcheries, and Taiwan is almost unique in the shrimp aquaculture industry in not relying on capture of wild seed. Another key development was the formulation of artificial feeds suitable for penaeid culture. From 1977 to 1986, the number of Taiwanese shrimp feed mills grew from three to over fifty with a total output of 100,000 metric tons per annum (U.S. Department of Commerce 1988, 1992a). To complete this chapter of our story and anticipate the next, due to land scarcity and high fixed costs, Taiwan pioneered intensive methods for shrimp production involving high stocking densities, aeration, and frequent water exchanges. Peak land use (in 1988) amounted to only 17,460 hectares, which is negligible when compared to Indonesia's 250,000 hectares largely under extensive production (U.S. Department of Commerce 1992a).

These Ecuadorean, Japanese, Hawaiian, and Taiwanese early actors built technical and business capacity for eventual private sector development of a pond-based industry which was complemented by the numerous feasibility studies, fisheries sector studies, and joint venture opportunities made available by international lending agencies such as the World Bank and the Asian Development Bank (cf. Aquatic Farms 1989). In addition, the role of national banks is noteworthy in the growth of the shrimp pond industry (e.g., Thailand's Bangkok Bank, Ecuador's Central Bank and Banco de Fomento). Attractive also was the impetus provided by Asian governmental policies and investment incentives favorable to the development of shrimp pond production (U.S. Department of Commerce 1988). Up to this point, however, it would be possible to substitute almost any high-value fisheries commodity (e.g., lobsters, abalone) into this scenario (cf. Pillay 1990, 499–510). Hence the question is: Why and how did shrimp pond culture emerge in such a rapid and dramatic fashion?

THE EMERGENCE OF GLOBAL POND PRODUCTION

This section reviews the confluence of political, economic, and technical forces within a global system that has given rise to the present state of global shrimp pond aquaculture. The political, economic, and technical forces involved in each national situation have been examined in greater detail in Harris and Skladany (1993). This section loosely follows Friedland's (1984, 221–35) five-stage methodology for commodity systems analysis, focusing on production factors and practices, production organization, labor, technoscientific base, and marketing and distribution networks. Our emphasis is on production since various actors have made use of technical breakthroughs that catalyzed widespread entry into coastal shrimp pond production.

Production Factors and Practices

Extensive shrimp pond production systems exhibit relatively low capital requirements and previously made use of existing capture fisheries infrastructure, markets, and distribution networks. These production systems reflected the lack of "modern" technoscience, wage labor inputs (Hannig 1988; *World Shrimp Farming* May/June 1993, 8–9), and vertical market integration. During the past two decades, stagnant or declining harvests of wild caught shrimp from the sea; increasing labor, capital, and energy costs; and continued liberal import policies and growing market demand from the United States, Japan, and Europe have stimulated attempts to increase production in ponds. As a consequence of the increased confluence of the necessary production factors, pond raising of shrimp is now able to supply over 30 percent of the global demand for shrimp (Table 9.3).

Semi-intensive systems require more capital investments (e.g., for specific pond design and construction), higher operating costs for purchased seed and feed, and further integration into the modern capitalist export market economy. Production outputs are correspondingly much higher in semi-intensive systems (about 1,000 to 2,000 kgs/ha/yr) than in extensive systems (about 200 to 500 kgs/ha/yr). Semi-intensive production systems also require greater labor and management skills than do extensive ones. Due to the relative level of technical complexity associated with semi-intensive production, factors such as seed supply, feed quality, and disease control are also important management considerations; therefore, external factors related to the development of hatcheries, feed mills, shrimp health, and information services are important infrastructural entities when considering the global shift from extensive to semi-intensive production (McVey and Hanfman 1991).

Table 9.3
Estimated Cultured Shrimp Harvests, 1987–1991

Region/ Country	1987	1988	Year 1989	1990	1991	Change 1987–91
			1,000 Metric Tons			Percent
ASIA						
China	153	199	175	150	145	-5
Indonesia	73	96	97	120	140	92
Thailand	24	56	94	100E	110E	358
India	22	25	25	30	35E	59
Vietnam	20	22	22	30E	30E	50
Philippines	36	45	48	54	30E	-16
Taiwan	115	44	32	30	30E	-74
Other	23	25	29	35	38	65
Subtotal.........	466	513	522	549	558	20
LATIN AMERICA						
Ecuador	69	70	64	70	100	45
Colombia	1	1	3	6	10	900
Mexico	4	6	5	8	9	125
Peru	2	2	4	5	6	200
Honduras	2	2	3	3	5	150
Other	5	8	8	10	9	80
Subtotal.........	83	89	87	102	139	67
AFRICA	1	1	1	NA	NA	-
MIDDLE EAST	Negl	Negl	Negl	Negl	Negl	-
EUROPE	Negl	Negl	Negl	Negl	Negl	-
NORTH AMERICA	1	1	1	1	2	100
World Total	551	604	611	652	699	27

E - Estimated
NA - Not Available
Negl - Negligible
SOURCE: United States Department of Commerce (1992a).

Riparian Sites

A mere fifteen years ago, coastal areas were largely considered wastelands with little economic potential (Bailey 1988b; Csavas 1988; U.S. Department of Commerce 1992b). Despite the significant levels of traditional resource exploitation in these critical coastal zones by local populations, formal property title and rights were rarely recognized by nation-states. This ambiguity concerning property regimes is the basis for the explosive growth of shrimp pond aquaculture. Transnational corporations (TNCs), national corporations (NCs), consultants, and elements of the state have taken great advantage of the natural resources found in the coastal zones by privatizing or appropriating them, and material production in these rural coastal lands manifests a set of characteristics and relations analogous to a ''hyper-intensive'' free trade zone. One result is that previously ''worthless'' swamps, forests, and marginal agricultural lands have become subject to capitalist encroachment and commodification. Land prices have risen accordingly, and speculation is rampant.

Seed

Due to the high demand for seed, the development of shrimp hatcheries has taken place at phenomenal levels. Hatchery production in China grew from 337 million seed (post larvae, or pl) in 1980 to over 74 billion pl in 1,000 hatcheries largely operated by the Chinese state (U.S. Department of Commerce 1992a). With the noted exception of Taiwan, hatcheries have yet to close the production cycle completely as gravid females taken by trawl fisheries are sold to hatcheries. In Ecuador the preference is for wild pl seed which are harvested by a reported 90,000 to 140,000 seed collectors along the coast (Meltzoff and LiPuma 1986; U.S. Department of Commerce 1992d). Nonetheless, it is possible to maturate female penaeid shrimp in captivity, and the cycle may eventually close but is not expected to in the foreseeable future.

Feed

As the global shift from extensive to semi-intensive production unfolds, numerous feed mills have developed formulated feeds for shrimp grown in the larger producing countries. In Thailand, the giant agro-industrial conglomerate Charoen Pokphand has entered into the regional shrimp-farming business in a major fashion. Modeled on the near monopoly of the commercial production of poultry, the Charoen Pokphand company operates the world's largest feed mill. Other countries in the region such as Burma, Vietnam, and Bangladesh possess no feed industry and are constrained to largely extensive production, involving some pond fertilization or reliance on imported feed. Charoen Pokphand, however, is preparing for massive investments in Vietnam and already operates in Indonesia and China (U.S. Department of Commerce 1992a). China's feed industry is restricted to combinations of trash fish and shellfish with supplements mixed at small local mills. While China's production costs remain among the world's lowest (at U.S. $2.00 per kilogram), the feed quality is considered poor. At present, China has entered into joint venture agreements with Charoen Pokphand and a Norwegian firm, Trouw International Company, to provide an array of agro-industrial inputs including feed for shrimp pond production (U.S. Department of Commerce 1992a).

Ecuador has twenty-five operating mills (54,500 tons in 1988) which produce inexpensive feed of varying quality. If the nascent Latin American shrimp industry is to continue to hold its market share in the global shrimp trade, it will have to move toward more semi-intensive and intensive operations. This will entail higher demand for feed while seeking to hold productions costs down. One way in which this can be accomplished is by developing open trade arrangements, for example with Chile and Peru, to take advantage of relatively cheap sources of fish meal. Other avenues, such as soybean substitute for fish-meal from Brazil, may become important once more is known about shrimp nutrition. At the same time, however, Ralston Purina is especially active in the region (U.S. Department of Commerce 1992b,c).

Other Inputs

It is with the shift from semi-intensive to *intensive* forms of production that other inputs join seed and feed in their importance for the production process. In intensive systems, high capital investment, high stocking densities, frequent water exchanges, mechanical aeration, high feed levels, skilled labor, and sci-entific management result in the highest and riskiest production levels currently attained. Many transnationals and national firms in Asia undertake intensive production. Taiwan, due to an already established technoscientific base, as well as limited and highly priced land, engaged in intensive production early in the 1980s. The industry grew rapidly in the mid-1980s, and Taiwan emerged as the second largest producer in 1987 with a recorded 115,000 metric tons of pond-cultured shrimp. Due to the close proximity of the ponds, the level of intensity deployed in the pond production process, an extensive chain of intermediate handlers of pl stock, and haphazard water exchanges into and out of ponds, shrimp producers were hit with a massive disease outbreak in 1988 that deci-mated the industry. It remains to be seen whether the Taiwanese industry will recover. Significantly, however, Taiwanese culturists have expanded into other Asian and Latin American countries bringing with them intensive methods of shrimp pond cultivation.

Production Organization

Over the past ten years, one barely discernable trend has emerged in the globalization of the shrimp pond industry: The largest firms are increasing direct and, to a greater extent, indirect control over all phases of the shrimp commodity chain (Chong 1990). Transnational giants, such as Mitsubishi, British Petroleum-Aquastar, Charoen Pokphand, Ralston Purina, and other firms, are consolidating market shares in numerous Asian and Latin American countries. Consulting firms are also heavily engaged in the shrimp-farming business. This trend toward concentration is illustrated in Table 9.4; using three firms as examples, it is evident that vertical integration is extensive. One of the ways in which control over production is accomplished is through contract culturing.

Table 9.4 also illustrates the trend of globalization; again, the three firms described show an extensive pattern of geographic dispersal of operations. In some cases, operations undertaken by Aquastar and Charoen Pokphand in south-ern Thailand integrate their operations from pl seed production and contract farming through to eventual market outlets in Japan, Europe, and the United States. Aquastar has offices in the United States and Europe to coordinate market

Table 9.4
Partial Profile of Three Transnational Firms Engaged in Shrimp Pond Development

Firm	HQuarters	Description	Clients/Countries
Charoen Pokhpand Group	Bangkok, THAILAND 13 offices world wide	**Large Agricultural Conglomerate;** primarily poultry; involved in all phases of shrimp aquaculture including -exports -feedmills -extension -contract farming -farm operations -seed stock -processing plants	Clients State Governments State Fisheries Ministries and Departments Private Sector Countries Thailand China Vietnam Indonesia Mexico
Aquatic Farms Ltd.	Honolulu, Hawaii USA	**Aquaculture Consulting;** all phases especially shrimp -feasibility & sector studies -joint ventures -troubleshooting -farm management -farm operations -training	Clients Asian Development Bank World Bank State Governments State Fisheries Ministries and Departments Private Sector Countries Thailand Burma Bangladesh Pakistan India Philippines Malaysia Latin America
Ralston Purina International (subsidiary of R. Purina Company)	St. Louis, Missouri USA	**Animal Feeds:** including shrimp feeds -exports -feedmills -extension -consulting -joint ventures -financing feed costs	Clients State Governments USAID Private Sector Countries Guatemala Mexico Colombia Brazil Peru Indonesia

*This is only a **partial** listing of TNC involvments
Sources: Aquatic Farms Limited 1988a; 1988b; U.S. Department of Commerce 1992a; 1992b; 1992c; 1992d; World Shrimp Farming 1991b: 6-7).

trade. Charoen Pokphand also operates thirteen offices worldwide to oversee its diverse export markets including shrimp (Petrocci 1992; U.S. Department of Commerce 1992a; *World Shrimp Farming* March/April 1993, 6). In addition, many Taiwanese groups are exporting both technology and associated input factors throughout the world.

Labor

Labor patterns in the growing shrimp culture industry are difficult to discern. In general, there seems to be an influx of unskilled labor from agriculture into daily operations of semi-intensive ponds. The intensive operations use more skilled labor which is relatively well paid when compared to wage labor in agriculture. The use of technical specialists and outside consultants is also evident on some of the larger commercial or contracted farms. In addition, local fisheries department officials and university academics in various countries often act as business partners or ''weekend'' consultants to various farms. The importance of international consulting firms cannot be overstated. Often based in Hawaii, firms such as Aquatic Farms have enjoyed a close business association with various donors, governments, corporations, investment groups, and individuals (Table 9.4). Strategies employed by these firms relate to production increases, troubleshooting, and economic and financial feasibility studies. In some cases, firms operate farms and often assist the setting up of joint venture facilities. Consultants travel widely and often have the opportunity to keep abreast on breaking field developments.

Technoscientific Base

Research in shrimp culture entails a close mix of academic, governmental, and private initiatives. As more countries enter into the global shrimp market, a concerted drive appears to be taking shape with respect to developing better and more cost-efficient feeds, hatchery capacity, and disease control (Sandifer et al. 1991; International Trade Centre 1983, 13). This situation varies by region (Vanderpool, Harris, and Merson 1987). Asian countries possess substantial technoscientific capability evidenced by the proliferation of consulting firms, fisheries agencies, and business organizations. In contrast, Latin American countries rely heavily on U.S. academics and consulting firms for their technical needs (McVey and Hanfman 1991). Little effort is expended by Latin American governments toward research in shrimp culture.[3]

Marketing and Distribution Networks

Up to this point, we have focused on the segment of the shrimp commodity chain concerned with the production of raw shrimp. At the same time, a global infrastructure has evolved around the postharvest phases (Pillay 1990, 244–52).

Table 9.5
Shrimp Imports by Major Markets, 1986–1989

Country	Year				
	1986	1987	1988	1989	1990
	1,000 Metric Tons				
United States	160.5	193.6	206.5	214.4	227.4
Japan	213.8	246.6	258.7	263.7	284.3
EC	112.0	122.8	154.6	172.0	188.1
Other*	109.9	118.8	152.1	156.8	NA
Total	596.2	681.8	771.9	806.9	856.6**

NA - Not available
* - includes Hong Kong, Singapore, Malaysia and Canada
** - 1989 other figure added to 1990
SOURCE: United States Department of Commerce 1992b

Over 80 percent of cultured shrimp enters into the global market, but this varies by region. Domestic markets in Asia consume 20 percent of regional shrimp production, whereas 95 percent of Latin American production is exported outside the region (Rosenberry 1990; World Shrimp Farming Annual 1992). The major markets for the consumption of pond-raised shrimp are in the United States, Japan, and the European Community (Table 9.5).

As can be seen in Table 9.5, import demand is growing in both the United States and the European Community, however, Japan's import market demand tends to exhibit instability (Yu 1987) and seems to be leveling off (Yoshinori 1987). In the 1980s, Japanese consumers began to accept pond-raised *Penaeus monodon* imports from Southeast Asia only to switch preferences to wild caught *Penaeus monodon* due to increased concerns over perceived health risks (U.S. Department of Commerce 1992b). This consumer perception stems from reports that indicate that Asian growers use large amounts of antibiotics in order to control disease problems. Japan's Ministry of Health also found high levels of antibiotics in Thai and Philippine shipments of cultured shrimp (U.S. Department of Commerce 1992b). As a consequence, Japanese traders have renewed their efforts to obtain wild caught shrimp.

Processing

Once shrimp are harvested they are transformed to a lesser or greater extent by processing operations. One strong tendency in the U.S. market is that substantial product differentiation is taking place away from the traditional "peel and eat" forms of consumption. New commodity forms such as analogs (molded mixtures of shrimp and fish—known as *surimi* in Japan), breaded shrimp, shrimp cocktails and attractive packaging forms exclaiming the virtues of pond-raised shrimp are fast growing segments of the international shrimp market (Thompson, Roberts, and Pawlyk 1985; U.S. Department of Commerce 1992a–d; Vondruska 1985).

In general, processing facilities in Latin America remain limited in nations with small shrimp capture and culture bases (e.g., Brazil, Venezuela, and French Guiana), but they are expanding rapidly in the major producing nations (e.g., Colombia, Ecuador, and Peru). In Central America, Mexico is considered to hold the greatest potential for pond culture of shrimp. The pending North American Free Trade Agreement is expected to stimulate massive capital investments and joint ventures between Mexican firms and various TNCs. The close proximity to the U.S. market also offers Mexico a potentially unique comparative export advantage.

At the early stages of processing, U.S. consumers may not clearly distinguish among species or product forms. Few U.S. consumers, for instance, differentiate between pond-cultured and wild caught shrimp (U.S. Department of Commerce 1992b). Further, the majority of shrimp are consumed in situations where the consumer has no direct control over its characteristics; for example, in the United States, over 70 percent of shrimp is consumed outside the home, in restaurants. In this situation, product differentiation is as much a creation of advertising as it is a reflection of traditional cultural values or intrinsic significant differences. Florida shrimp firms have begun a campaign to develop a market identity for Florida shrimp, which currently are almost entirely wild caught. In contrast, SeaPak assures restaurant operators that "Thanks to shrimp farming freshness has taken on a whole new meaning at SeaPak. And words like control and consistency have been added to the shrimp industry vernacular" (from an advertisement in *Restaurant Hospitality*, May 1993).

GLOBAL RESTRUCTURING OF THE SHRIMP INDUSTRY

In viewing the global emergence of shrimp pond culture, we note a tendency to intensify operations at every level of the commodity chain. Internally, within a specific nation-state, efforts are being made to provide growers with auxiliary services including seed stock, formulated feeds, extension information, and marketing outlets. Firms have undertaken efforts to contract farmers by guaranteeing support and prices in return for a standardized product output. The increased horizontal and vertical integration of shrimp farming in the large producer countries leads to relatively high concentration indices. In Thailand, a global shrimp culture production leader, the Charoen Pokphand Company, dominates shrimp feed production with a 70 percent market share. Nine other feed mills hold from 1 to 4 percent of the market each (*World Shrimp Farming* May/June 1992, 3–4). These firms integrate production through postharvest activities and either own large amounts of land or indirectly control production through contract farming (Chong 1990).

In conjunction with concentration within the industry, firms are also expanding operations into new environments, what Lin (1992) calls "slash and burn." Again, the example of the Charoen Pokphand company in Thailand shows that substantial investments in China, Indonesia, and Vietnam are currently under way. In addition, Charoen Pokphand is exploring opportunities in Mexico (U.S. Department of Commerce 1992a; *World Shrimp Farming* May/June 1992, 3–4, 1993b, 5–6). The opening up of Mexico's coastline and close proximity to the United States make Mexico an ideal area for the expansion of shrimp culture. As noted earlier, Taiwanese technology has also played a major role in the expansion of the shrimp culture industry. Currently, two large Taiwanese intensive farms are operating in Texas although with mixed success (*World Shrimp Farming* January/February 1993, 2–10). Throughout the world, Taiwanese capital and technology are currently active in such countries as Thailand, China, the United States, Belize, the Dominican Republic, Panama, Colombia, El Salvador, Guatemala, Saint Lucia and Saint Vincent (U.S. Department of Commerce 1992b; *World Shrimp Farming* May/June 1993, 22–23). U.S. interests, especially corporations and consultants, are also very attractive worldwide.

What both concentration and expansion mean within the world shrimp commodity system is that corporate control and organizational forms will probably emerge to define and dominate the industry much as in the poultry and vegetable commodities (Heffernan 1984; Wilson 1985). As new strains of shrimp are developed to resist diseases, and to respond to company feed and antibiotics, firms will consolidate control over the production regime. Firms can quickly create the type of local infrastructure needed in those countries that do not have it. In Indonesia, private firms built roads and installed electricity in previously remote regions in order to bring shrimp to market terminals. In addition, for example, Charoen Pokphand's extension service, which includes mobile, fully equipped disease diagnosis laboratories, offers producers an immediate response in contrast to slower state services. In essence, pond culture shrimp involves a transition from a relatively specific capture-to-market basis to global production to a world consumption system. As a result, the role of the nation-state is seen as becoming intermediate owing to the increased involvement of multinational firms and international agencies and the requisite global infrastructure at all levels of the commodity chain (Bonanno 1991).

SHRIMP IN THE GLOBAL POLITICAL ECONOMY: SOCIAL AND ENVIRONMENTAL JUSTICE FOR WHOM?

A key factor underlying the emergence of the global shrimp commodity system has been the transformation of property and property rights (Meltzoff and LiPuma 1986; Bailey 1988b; Pillay 1992, 32–38). In conjunction with high demand and prices in international markets, the ease with which TNCs, NCs, and the state have transformed multipurpose, multiuser public coastal land into single-purpose private property has been the key factor underlying the rapid expansion and transformation of the shrimp pond culture industry. Expropriation of coastal residents is often undertaken through coercive measures. Coastal residents who have resided in a particular area for generations face the threats of ''modern'' state-sanctioned legal action and, in some cases, even murder (Bailey 1988b; Eckachai 1988; Skladany 1992; Srisuksai 1990; *World Shrimp Farming* January/February 1992, 4, November/December 1992, 7, January/February 1993, 8–9, May/June 1993, 8–9, 23). In response to these turbulent instances, the state is called upon to fulfill its functions of legitimation and regulation; the call eventually centers around the need for some form of coastal zone management (CZM). Both Ecuador and Thailand offer interesting cases that clarify the incapacity of the state underlying rapid conversion of coastal land and mangroves into shrimp ponds by TNCs and NCs.

These cases also illustrate the ''myth'' of CZM and the increasingly fragmented and incapacitated role of the state. In the 1980s, both governments have formally become increasingly involved in attempts to regulate the shrimp pond industry. Through a cabinet-level National Mangrove Committee, Thailand enacted a proposed mangrove and coastal land classification scheme (Sirisup 1988, 66–9). Coastal land was zoned into three categories—conservation and two economic categories. Practices undertaken in the actual zones are widely ignored, and the scheme relied on information from 1969. In Chantaburi province, where Sirisup (1988) undertook one of the few social studies made of this particular industry, pond owners claimed that they did not know where the boundaries of the particular zones were.

In the 1970s and early 1980s, obtaining a ''free'' concession permit to build a shrimp pond in Ecuador involved having access to the proper government channels who controlled coastal land allocation. Meltzoff and LiPuma (1986) stated that quick approval can be obtained by a series of payments to various government officers in charge of permit authorization; as of 1986, payments were in the range of U.S. $10,000 per 100 hectare concession. In fact, what has often been the case is that government officials, bank loan officers, and businessmen form partnerships. This arrangement speeds up the process of obtaining a concession if undertaken by high-level government of military officers.

It is ironic that the rapid property transformations that underlie shrimp pond development were contemporaneous in both Thailand and Ecuador with internationally sponsored CZM projects. While providing some profiles of coastal zones and appeals to First World planning and enforcement procedures, these projects were largely ineffective in understanding the expansion of the industry let alone in proposing ideas on how to regulate it. Merschod (1989) advocates a *compradrazgo* participatory process involving all parties who hold a stake in the coastal zone. Bringing shrimp pond organizations into CZM policy dialogue offers, on the surface, a seemingly rational course of action. In fact, these constituencies can use CZM proposals and laws to eliminate new entrants and competition into the coastal zone.

ENVIRONMENTAL AND SOCIAL CONSEQUENCES

Negative environmental and social consequences are recognized in the literature pertaining to the shrimp pond industry (Bailey 1988b, 1992; Bailey and Skladany 1992; Chong 1990; Eckachai 1988; Macintosh 1982; Macintosh and Phillips 1992; McCarthey 1990; Meltzoff and LiPuma 1986; Pullin 1989; Sirisup 1988; Skladany 1992; Stonich 1992; U.S. Department of Commerce 1988, 1992a–d; Weeks 1992). This literature is minuscule, however, when compared with the pro-business orientation and advocacy position in aquaculture journals, magazines, reports, and trade publications (e.g., Aquatic Farms Limited 1989; Bashirullah, Mahmood, and Matin 1989; Csavas 1988, 1990a–c; Petrocci 1992; Scura 1985, *World Shrimp Farming* passim). Social scientists (excluding economists) for the most part are outsiders in terms of the industry, and no widely disseminated and concerted study is currently attempting to investigate the environmental and social consequences of shrimp farming. In contrast, businesspersons, aid officials, and consultants, among others, repeat a litany of perceived beneficial developments (e.g., foreign exchange earnings, jobs, skills training, sustaining the small-scale farm) associated with this particular industrial production system (Skladany 1992).

Negative and unregulated environmental and social impacts are widely associated with shrimp pond development both in mangrove areas and in adjoining coastal land (Bailey 1988b; Meltzoff and LiPuma 1986; Stonich 1992). The key factor is that transformation of property regimes allows for construction of private ponds and emplaces relatively irreversible structures in environmentally sensitive areas—the critical point where the sea meets land. Expansion of the culture of marine shrimp such as *Penaeus spp.* into agricultural lands requires that salt water be pumped from the sea. Discharge of pond water can have the effect of turning adjacent paddy land into saline areas which are no longer suitable for rice cultivation. Again, spokespersons for firms tend to place the blame on "wildcatters" in contrast to the "environmentally conscientious" image put forth on the part of the "green" TNC. In fact, recent evidence suggests that these firms are highly mobile and are quite willing to relocate when conditions deteriorate. New frontiers are currently being opened in countries such as Muanmar, Brazil, and Mexico and elsewhere including Africa.

This mobility gives rise to the pattern called "slash-and-burn" development. Parodying the conventional view of swidden agriculture, slash-and-burn development refers to the destruction and extraction of environmental values without concern for conservation or regeneration for sustainable production (Bunker 1985). In the case of shrimp culture, one can see this in Southeast Asia. When Taiwan's shrimp culture industry was decimated by disease and land subsidence in 1987 (Liu and Tu 1988), the industry relocated to new countries, leaving behind an extensive area where aquaculture was no longer possible. Taiwanese shrimp culture investors and specialists worked closely with Thai, Indonesia, and other shrimp culture agents to develop the industry in these countries. But when Thai production began leveling off, Taiwanese interests moved on to Texas and Central and Latin America. The irony of this process is that, in contrast to traditional swidden agriculture where slash-and-burn techniques are one phase of a longer sustainable production cycle, slash-and-burn shrimp farming leaves the land unsuited to aquaculture or agriculture for at least several centuries.

CONCLUSIONS

The social and environmental problems of shrimp pond production are international in scope and formulation, regional in specific manifestation, and local in terms of social impacts (Skladany 1992). The boom-and-bust cycle of slash and burn, the increased concentration and expansion of TNC operations, and the elite class expropriation of marginal classes from livelihoods and natural resources have all led to substantial, long-term, negative social consequences which will only become more exacerbated in the foreseeable future. In short, the industry is not accountable and seeks short-term gains by sidestepping issues of long-term environmental degradation and social justice. In essence, a significant structural, national, and international transformation is taking place in a series of regions which articulate into a global commodity regime. It is the TNCs and NCs that direct the shape and trajectories of this new global regime.

Increasingly, the role of the state can be viewed as a mere backdrop against which rapid capital flows, international finance, technoscientific sophistry, environmental degradation, and social injustice are perpetuated by highly mobile TNCs and NCs. By providing initial investment incentives and generous aid, states and international development agencies have set into motion a series of contradictions that have reset the limits in terms of state capacity for capital accumulation and social legitimation. The two principal contradictions are (1) the attempt to base sustainable development on the superexploitation of natural resources and (2) the reliance on TNCs as the vehicle of development even though the factors of production that they bring to the process are highly mobile. Nowhere does the diminution of state capacity become more apparent than in property transformations underlying the explosive growth of the shrimp pond culture industry. Unlike other world agricultural and food systems, intensive capitalist identification and commodification of the land in coastal zones have been very recent. As soon as a series of sociohistorical factors involving an exhausted shrimp capture base, technical advances, international aid, growing markets, liberal state trade and investment policies, institutional support, and public and private incentives became apparent and coincided with elite class interests, TNCs and NCs were uniquely positioned to transform property and natural resources into a global industry at great social costs.

The ability of the nation-state to manage these contradictions effectively is simply overwhelmed by the international scope of the industry's fluid structure, which represents neither region nor context. There is, at present, no foreseeable effective strategy to deal with the negative consequences associated with shrimp culture. In addition, the emergence of a global shrimp aquaculture industry may be a harbinger of future trajectories in the commodification of aquatic organisms on a global level. At the same time, the continued development of shrimp pond culture creates a series of contradictions characterized by its mode of production, its markets, and its distribution networks. In essence, the human dimensions of social and environmental justice and injustice remain unimproved.

NOTES

This is a revised version of a paper presented at the seventeenth annual Political Economy of the World-System Conference, held at Cornell University, Ithaca, New York, April 17, 1993. The authors would like to express their appreciation to Philip McMichael and Wilfred Harris for helpful comments on an earlier version.

1. Pilot farm efforts are also under way in several African nations (e.g., Cameroon, The Gambia, Ivory Coast, Kenya, Madagascar, Senegal, Seychelles, and Zambia). Africa is considered to have substantial potential for pond raising of shrimp, but production is expected to be negligible as a part of world production in the foreseeable future (U.S. Department of Commerce 1992a).

2. Interestingly, these figures are disputed by the Ecuadorean Chamber of Shrimp Producers who claimed in 1989 that only 33,000 hectares of mangroves were destroyed by shrimp farmers, charcoal makers, and coastal developers (U.S. Department of Commerce 1992d, 888). Regardless of the actual figures, what we see emerging in Ecuador as well as elsewhere are attempts to blur the understanding of the extent of environmental degradation, to sidestep social and environmental accountability, and to deflect from issues concerning environmental and social justice.

3. A few examples will suffice to illustrate. Taiwan's Fisheries Research Institute developed pivotal seed production, feeding regimes and intensive pond cultivation through the mid-1980s. Although the government no longer encourages shrimp production, substantial private initiative exists such that Taiwanese expertise and capital have found second country applications throughout the world. This trend is likely to continue in the Asian region as well as in other south-to-south contexts and even in south-to-north situations; several operations on the Texas Gulf coast are owned by Taiwanese firms (*World Shrimp Farming* January/February 1993, 2–11).

China supports a strong network of research institutes supervised by the Chinese Academy of Sciences. A variety of research is conducted at these institutes including identification of new species, disease identification and control, selective breeding, and increased production techniques. The results of this network have elevated China into the top position as a producer of pond-cultured shrimp.

In Thailand, the Royal Thai Department of Fisheries operates a provincial network of brackish water fisheries and research stations which were built largely with loans and grants from a variety of bilateral and multilateral aid agencies. The Thai TNC Charoen Pokphand operates fourteen shrimp culture extension centers each well equipped with laboratories.

In Latin America, the government tends not to support research on shrimp culture. As a result, many Latin American firms make use of outside consultants from the United

States. Often a joint venture operation will hire and employ U.S. scientists and consultants to conduct applied research on the company's farm.

REFERENCES

Anderson, Benedict. 1990. "Murder and Progress in Modern Siam." *New Left Review* 181: 33–48.

Aquatic Farms Limited. 1988a. Newsletter. July, 8 pp.

————. 1988b. Newsletter. December, 8 pp.

Aquatic Farms Limited. 1989. "Asia-Wide Shrimp Agro-Industry Sector Study." Report submitted to the World Bank. June 1989. Honolulu, Hawaii: Aquatic Farms, Ltd.

Bailey, Conner. 1988a. "The Political Economy of Fisheries Development in the Third World." *Agriculture and Human Values* 5, 1–2: 35–48.

————. 1988b. "The Social Consequences of Tropical Shrimp Mariculture Development." *Ocean & Shoreline Management* 11: 31–44.

————. 1992. "Coastal Aquaculture Development in Indonesia." In *Contributions to Fishery Development Policy in Indonesia*, edited by Richard B. Pollnac, Conner Bailey, and Alie Poernomo. Jakarta, Indonesia: Central Research Institute for Fisheries. Agency for Agricultural Research and Development, Ministry of Agriculture.

Bailey, Conner, and Mike Skladany. 1991. "Aquacultural Development in Tropical Asia: A Re-evaluation." *Natural Resources Forum* 15, 1: 66–74.

Bailey, Conner, A. Dwiponggo, and F. Marahudin. 1987. "Indonesian Marine Capture Fisheries." *ICLARM Studies and Reviews* 10. Manila, Philippines: International Center for Living Aquatic Resources Management.

Bashirullah, A.K.M., N. Mahmood, and A.K.M.A. Matin. 1989. "Aquaculture and Coastal Zone Management in Bangladesh." *Coastal Management* 17: 119–27.

Baud, I.S.A. 1992. *Forms of Production and Women's Labour*. New Delhi: Sage Publications.

Bonanno, Alessandro. 1991. "The Globalization of the Agricultural and Food System and Theories of the State." *International Journal of Sociology of Agriculture and Food* 1: 15–30.

Borgstrom, Georg. 1967. *The Hungry Planet*. New York: Collier Books.

Brown, E. Evan. 1989. *World Fish Farming: Cultivation and Economics*. Westport, Conn.: AVI Publishing, 381–86.

Bunker, Stephen G. 1985. *Underdeveloping the Amazon*. Chicago: University of Chicago Press.

Chaston, I. 1983. *Marketing in Fisheries and Aquaculture*. Oxford: Fishing News Books.

Chen, Tung Pai. 1976. *Aquaculture Practices in Taiwan*. Farnham: Fishing News Books.

Chong, Kee-Chai. 1990. "Asian Shrimp Aquaculture—At Crossroads." *Infofish International* 5, 90: 40–47.

Csavas, Imre. 1988. "Shrimp Farming Development in Asia." Paper presented at Shrimp '88, INFOFISH, Kuala Lumpur, Malaysia, June.

————. 1990a. "Shrimp Aquaculture Developments in Asia." In *Technical and Economic Aspects of Shrimp Farming*, edited by Michael B. New, Henri de Saram, and Tarlochan Singh. Proceedings of the Aquatech '90 Conference, Kuala Lumpur, Malaysia, June 11–14.

————. 1990b. "New Developments in Asian Aquaculture." Paper presented at the Aquaculture International Congress and Exposition, Vancouver, Canada, September 4–7.

————. 1990c. "Aquaculture Development and Developing Countries of Asia." Paper presented at the Conference on Environmental Issues in Third World Aquaculture Development, Bellagio, Italy, September 17–22.

Eckachai, Sanitsuda. 1988. *Behind the Smile: Everyday Faces of Thailand.* Bangkok, Thailand: Duang Kamol.

Food and Agriculture Organization of the United Nations. 1989. "Review of the State of World Fishery Resources." FAO Fisheries Circular no. 710, Revision 6. Rome: FAO.

Friedland, William. 1984. "Commodity Systems Analysis: An Approach to the Sociology of Agriculture." In *Research in Rural Sociology and Development,* edited by H. K. Schwarzweller. vol. 1. Greenwich, Conn.: JAI Press.

Goonatilake, Susantha. 1984. *Aborted Discovery: Science & Creativity in the Third World.* London: Zed Books.

Gulland, John A., and Brian F. Rothschild. 1984. "Preface." *Penaeid Shrimps—Their Biology and Management.* Surrey, England: Fishing News Book Limited.

Hannig, Wolfgang. 1988. *Towards a Blue Revolution: Socioeconomic Aspects of Brackishwater Pond Culture in Java.* Jogjakarta, Indonesia: Gadjah Mada University Press.

Hanson, Joe A., John E. Huguenin, Suzzane S. Huguenin, and Harold L. Goodwin. 1977. *Shrimp and Prawn Farming in the Western Hemisphere.* Stroudsburg, Pa.: Dowden, Hutchinson and Ross, 7.

Harris, Craig K., and Michael Skladany. 1993. "From Shrimp Boats to Prawn Ponds: The Rise of the Global Shrimp Aquaculture Industry." Department of Sociology working paper. Michigan State University.

Heffernan, William D. 1984. "Constraints in the U.S. Poultry Industry." *Rural Sociology* 49, 1: 237–60.

International Trade Centre, UNCTAD/GATT. 1983. *Shrimps: A Survey of the World Market.* Geneva: UNCTAD.

Klausen, Arne M. 1968. *Kerala Fishermen and the Indo-Norwegian Pilot Project.* Oslo: Universitetsforlaget.

Lee, D. O'C., and J. F. Wickens. 1992. *Crustacean Farming.* New York: Halsted Press, an imprint of John Wiley & Sons, 20–52.

Lin, C. Kwei. 1992. "Intensive Marine Shrimp Culture in Thailand: Success and Failure." Paper presented at Aquaculture '92, Orlando, Florida, May 24.

Ling, Shao-wen. 1977. *Aquaculture in Southeast Asia: A Historical Overview.* Seattle: University of Washington Press.

Liu, Yung-ching, and Tu Su-hao. 1988. *Environmental Impacts of Shrimp Ponds in Pingtung County, Taiwan.* Taipei: National Taiwan University Department of Agricultural Extension.

McCarthy, Florence E. 1990. "The Role of Foreign Assistance and Commercial Interests in the Exploitation of the Sundarbans." *Agriculture and Human Values* 7, 2: 52–60.

Macintosh, D. J. 1982. "Fisheries and Aquaculture: Significance of Mangrove Swamps." In *Recent Advances in Aquaculture,* edited by J. F. Muir and R. J. Roberts. Boulder, Colo.: Westview Press.

Macintosh, D. J., and M. J. Phillips. 1992. "Environmental Considerations in Shrimp Farming." *INFOFISH International*, June 1992, 3–82.

McVey, James P., and Deborah T. Hanfman. 1991. "Information Services in Aquaculture." In *Frontiers of Shrimp Research*, edited by P. F. DeLoach, W. J. Dougherty, and M. A. Davidson, 257–80. Amsterdam: Elsevier.

Meltzoff, Sarah Keene, and Edward LiPuma. 1986. "The Social and Political Economy of Coastal Zone Management: Shrimp Mariculture in Ecuador." *Coastal Zone Management Journal* 14, 4: 349–80.

Merschod, Kris. 1989. "In Search of a Strategy for Coastal Zone Management in the Third World: Notes From Ecuador." *Coastal Management* 17: 63–74.

Panayoutou, Theodore, and Songpol Jetanavanich. 1987. "The Economics and Management of Thai Marine Fisheries." ICLARM Studies and Reviews no. 14. Manila, Philippines: International Center for Living Aquatic Resources Management.

Paw, James, N., S. Bunpapong, A. White, M. Sol, and M. Sadorra. 1988. "The Coastal Environmental Profile of Ban Don Bay and Phangnga Bay, Thailand." Technical Publication no. 2. ASEAN/United States Coastal Resources Management Project. Manila, Philippines: International Center for Living Aquatic Resources Management.

Petrocci, Charles. 1992. "Aquastar—Partners in Progress for Aquaculture." *Aquaculture Magazine*, November/December, 30–36.

Pillay, T.V.R. 1990. *Aquaculture: Principles and Practices*. Oxford: Fishing News Books.

———. 1992. *Aquaculture and the Environment*. New York: Halsted Press.

Pullin, Roger. 1989. "Third World Aquaculture and the Environment." *NAGA: The ICLARM Quarterly* 12, 1: 10–13.

Rosenberry, Bob. 1990. "World Shrimp Farming: Can the Western Hemisphere Compete with the Eastern?" *Aquaculture Magazine*, September/October, 60–64.

Rosendal, Kristin. 1992. "Blue Revolution Could Avoid Failures of Green Predecessor." *Biotechnology and Development Monitor* 12 (September): 10.

Saito, Etsuo, Richard S. Johnston, and Arthur Siaway. 1985. "The Japanese Market for Shrimp." In *Proceedings of the Workshop on Shrimp and Prawn Markets*. Sponsored by the International Institute of Fisheries Economics and Trade and the South Carolina Wildlife and Marine Resources Department. Charleston, South Carolina, July 26–27, 33–38.

Sandifer, Paul A., J. Stephen Hopkins, Alvin D. Stokes, and Gary D. Pruder. 1991. "Technological Advances in Intensive Pond Culture of Shrimp in the United States." In *Frontiers of Shrimp Research*, edited by P. F. DeLoach, W. J. Dougherty, and M. A. Davidson, 241–56. Amsterdam: Elsevier.

Sarig, Shmuel. 1988. "The Development of Polyculture in Israel: A Model of Intensification." In *Intensive Fish Farming*, edited by C. J. Shepherd and N. Bromage, 333–64. Oxford, England: BSP Professional Books.

Scura, Edward D. 1985. "The Challenge and Potential of Aquaculture." Keynote address, 16th Annual Meeting of the World Mariculture Society, Orlando, Florida, January 13–17.

Shigeno, Kunihiko. 1979. *Problems in Prawn Culture*. Rotterdam: A. A. Balkena.

Sirisup, Siriluck. 1988. "Socio-economic Changes in the Course of Commercialization of Coastal Aquaculture in Chanthaburi Province, Thailand." Master's thesis, Asian Institute of Technology, Bangkok, Thailand.

Skladany, Mike. 1992. "Use Conflicts in Southeast Asia: An Institutionalist Perspective." *World Aquaculture* 23, 2: 33–35.

Srisuksai, Charoeinkij. 1990. "Prawn Farmers, Rice Farmers at Odds in South." *The Nation*, November 4.

Stonich, Susan C. 1992. "Struggling with Honduran Poverty: The Environmental Consequences of Natural Resource-Based Development and Rural Transformations." *World Development* 20: 385–99.

Taisuke, Miyauchi. 1987. "Last Stage of the Journey: The Distribution System for Frozen Shrimp." *AMPO* 18, 4: 44–52.

Thompson, Mark E., Kenneth J. Roberts, and Perry W. Pawlyk. 1985. "Structure Changes in U.S. Shrimp Markets." In *Proceedings of the Workshop on Shrimp and Prawn Markets*. Sponsored by the International Institute of Fisheries Economics and Trade and the South Carolina Wildlife and Marine Resources Department. Charleston, South Carolina. July 26–27, 11–24.

U.S. Department of Commerce. 1988. *Aquaculture and Capture Fisheries: Impacts in U.S. Seafood Markets*. Report prepared pursuant to the National Aquaculture Improvement Act of 1985. U.S. Department of Commerce. National Oceanic and Atmospheric Administration. National Marine Fisheries Service. Washington, D.C.: U.S. Government Printing Office.

———. 1992a. *World Shrimp Culture*. Vol. 1, *Africa, Asia, Europe, Middle East, North America*. Prepared by the Office of International Affairs. United States Department of Commerce. National Oceanic and Atmospheric Administration. National Marine Fisheries Service. Washington, D.C.: U.S. Government Printing Office.

———. 1992b. *World Shrimp Culture*. Vol. 2, Pt. 1, *Latin America Overview and Caribbean*. Prepared by the Office of International Affairs. United States Department of Commerce. National Oceanic and Atmospheric Administration. National Marine Fisheries Service. Washington, D.C.: U.S. Government Printing Office.

———. 1992c. *World Shrimp Culture*. Vol. 2, Pt. 2, *Central America*. Prepared by the Office of International Affairs. United States Department of Commerce. National Oceanic and Atmospheric Administration. National Marine Fisheries Service. Washington, D.C.: U.S. Government Printing Office.

———. 1992d. *World Shrimp Culture*, Vol. 2, Pt. 3, *South America*. Prepared by the Office of International Affairs. United States Department of Commerce. National Oceanic and Atmospheric Administration. National Marine Fisheries Service. Washington, D.C.: U.S. Government Printing Office.

Vanderpool, Christopher K., C. K. Harris, and N. S. Merson. 1987. "Marine Resource Development and Developing Countries: Rough Seas to Neptune's Treasure." In *The Right to Food: Technology, Policy and Third World Agriculture*, edited by Philip Ehrensaft and F. Knelman. Montreal, Quebec: Canadian Associates of Ben-Gurion University.

Vondruska, John. 1985. "Economic Assessment of Trends in Shrimp Fisheries, Markets, Mariculture and Analog Products." In *Proceedings of the Workshop on Shrimp and Prawn Markets*. Sponsored by the International Institute of Fisheries Economics and Trade and the South Carolina Wildlife and Marine Resources Department. Charleston, South Carolina, July 26–27, 77–98.

Weeks, Priscilla. 1992. "Fish and People: Aquaculture and the Social Sciences." *Society and Natural Resources* 5: 345–57.

Wilson, James. 1985. "The Political Economy of Contract Farming." *Review of Radical Political Economics* 18, 4: 47–70.

World Shrimp Farming Annual. 1989. Annual report. Published by Aquaculture Digest, 21 pp.

World Shrimp Farming Annual. 1992. Annual report. Published by Aquaculture Digest, 33 pp.

Yoshinori, Murai. 1987. "The Life and Times of the Shrimp: From Third World Seas to Japanese Tables." *AMPO* 18, 4: 2–9.

Yu, Mizuno. 1987. "History: How We Got Our Appetite for Shrimp." *AMPO* 18, 4: 33–41.

PART IV

Recomposition of Global and Regional Agro-Food Systems

10

Industrial Restructuring and Agrarian Change: The Greening of Singapore

Frances M. Ufkes

In the mid-1980s, the Singapore state initiated a comprehensive program to restructure its domestic agricultural sector. Although land commands a premium in the small city-state, for several decades, traditional, urban farming systems have been able to supply a significant share of the total national consumption of major wage foods such as pork, poultry, vegetables, and eggs. Agricultural restructuring involved a shift toward the production of higher value-added agricultural and horticultural commodities, both for domestic use and for export, a greater spatial concentration of agricultural production, and a recomposition of Singapore's food import dependence. Traditional farming systems were phased out by government mandate, and agricultural production was restricted to state-sponsored agrotechnology parks.

These efforts are indicative of a broader movement by the state to reposition the national economy within the currents of global capitalist restructuring in order to pursue new avenues of accumulation. The 1986 recession marked an end to twenty years of high growth, fostering revisions in Singapore's economic and industrial policy. During the 1970s and early 1980s, Singapore's economic growth, based upon an open trade and regulatory regime and foreign investment in export-oriented manufacturing, linked the interests of a corporatist state with the competitive pressures facing industrial capital in core countries. However, in an era of saturated markets, instability of international commercial trade, emergent regional trading blocs, and intense intercapitalist rivalries over world market share, the continued viability of this industrial regime was in question. In the mid-1980s, Singapore's industrial policy was redirected toward promoting high-value manufacturing and services industries.

Transformations in land use, or the landscape of accumulation as Singapore enters the next lap of economic growth, mirror these shifts in industrial and agricultural policy and represent processes of greening at both material and ideological levels. Materially, this involves the concrete transformation of the urban and regional form into one that is physically cleaner, greener, and more tropical- or island-like, characteristics perceived to be attractive to new international labor and capital flows. Delimited to state-sponsored agrotechnology parks, agriculture in Singapore has become increasingly sterile and physically compartmentalized from other arenas of social and economic life. Ideological greening processes undergird these physical transformations and efforts by the state to respond to the exigencies of world capitalist restructuring.

This chapter examines the role of the Singapore state in altering domestic social structures of food production and consumption to pursue new avenues of accumulation in the wake of post-1970s world capitalist restructuring. Efforts to transform domestic agriculture reflect the changing imperatives of national economic policy and planning as constrained by the crises and contradictions of world capitalist accumulation. Agricultural restructuring has also led to a recomposition of Singapore's food import dependence, thus generating changes in the nature of Singapore's international food relations. Of central concern is the interplay of state and capital in the creation of a regional supply network to source the mass-market foods no longer produced domestically. Therefore, the other side of Singapore's greening is the expansion of Fordist agro-export systems within the Southeast Asian periphery.

This chapter first examines the dynamics of Singapore's industrialization and evolving matrix of domestic food needs in world-historical perspective, within the conceptualization of food regimes (Friedmann 1982, 1991; Friedmann and McMichael 1989). Within food regime theory, domestic industrial restructuring and agrarian change are viewed as complex, interrelated processes, connected to, and driven by, imperatives within the international political economy. Accordingly, this chapter departs from the voluminous literature on Singapore's industrialization, in which relations of food production and consumption are largely ignored, and in which the state is commonly viewed in neoclassical economic or technocratic terms—as a market facilitator (e.g., Ariff and Hill 1985; Lim 1983); it also departs from the studies that fail to link changes in Singaporean agriculture to the currents of *both* domestic and world capitalist accumulation. The second section of the chapter focuses on post-1985 agricultural restructuring, emphasizing the banning of domestic pig production and the establishment of agrotechnology parks. The chapter concludes with a discussion of Singapore's changing food import dependence. Singapore now relies entirely upon imports to source fresh pork consumption needs; this dependence is central to the expansion of pig export systems in Thailand, Malaysia, and Indonesia, and the creation of new regulatory structures in Singapore to monitor the pig trade and stabilize pork prices.

FOOD REGIMES AND SOUTHEAST ASIAN INDUSTRIALIZATION

The conceptualization of food regimes as elaborated by Friedmann (1982, 1991) and Friedmann and McMichael (1989) links international relations of food production and consumption to dynamics of the multistate system and forms of accumulation characteristic of particular historical junctures. Friedmann (1982, 1991) and Friedmann and McMichael (1989) have described two world food regimes, and McMichael (1992) has presented the main elements constituting a new, third food regime emerging within post-1973 world capitalist restructuring. The first food regime (ca. 1880–1914), based upon British hegemony, integrated European industrialization with extensive agricultural systems in the settler colonies and states and colonial plantation agriculture in the tropics. The second food regime (ca. 1945–1973) was based upon U.S. hegemony, intensive accumulation in industry and agriculture, extension of the state system to former colonies, and forms of national and international regulation, such as Bretton Woods and the General Agreement on Tariffs and Trade (GATT), which anchored world market stability. Nation-states were extensively involved in social and economic regulation, exemplified by Fordism in the United States and Europe (Aglietta 1979; Lipietz 1987) and state-led industrialization in the Third World (McMichael and Myhre 1991) subsidized by cheap food disposed of via the US PL 480 food aid program.

Southeast Asian industrialization occurred within the rise and fall of the second food regime, framed by post–World War II Asian nationalism and Cold War geopolitics. The first two decades following World War II were characterized by heightened U.S. security interests in the region, prompted by the Communist takeover in China in 1949, the Korean conflict of 1950–1953, and nationalist movements in Indochina and Indonesia—all of which led to a flurry of U.S. economic and military commitments aimed at Soviet and Sino-Soviet containment (Dulles 1950; Fifield 1973; Kodras 1993). Within Pax Americana, the political and economic security of the Asia-Pacific region was anchored in the successful reconstruction of Japanese capitalism and its integration with the markets and raw materials of Southeast Asia (McMichael 1987). The late 1940s and 1950s, therefore, were a transitional period for Southeast Asia, as old imperialisms (British, Dutch, and French) waned and as the "new imperialism of the United States tested the ground of the rich but volatile domain south of China" (Buchanan 1972, 17).

The expansion of world trade during the 1960s and the emergent crisis of U.S. Fordism was also central to the rise of the Southeast Asian newly industrializing countries (NICs). Cheap food, in conjunction with development loans, foreign direct investment, export credits, and extensive national economic and social regulation, led to the creation of comparative advantage in a narrow range of labor-intensive manufactures by the East and Southeast Asian NICs for sale in U.S., European, and Japanese markets (Hamilton 1983; Landsberg 1979).

American transnational corporations (TNCs), and later, European and Japanese TNCs, relocated labor-intensive parts of the production process in industries such as textiles and electronics in these countries to reduce costs and remain competitive in home markets. State-led, export-oriented industrialization (EOI) in the Third World, thereby resolved, at least temporarily, the competitive crises facing industrial capital in core countries. Beginning in the mid-1960s, "US capitalists could not successfully restore profit margins through continued price increases" (Landsberg 1979, 57), and economic growth in the Southeast Asian NICs was achieved by integrating their industrial bases "in the needs and logic of the international capitalist economy" (1979, 54).

By the mid-1970s, the cohesive elements binding the second food regime were in disarray. The breakdown of the Bretton Woods system of currency regulation and related energy-food shocks of the early 1970s ushered in widely fluctuating international financial and commodity flows, intense intercapitalist rivalries over world market share, and heightened transnationalization within the agro-food system linked to new production relations, class relations, and consumption patterns in the world-economy (McMichael 1992; Friedmann 1993). Economic restructuring involves the renegotiation of relations among labor, capital, and the state reflected in post-Fordist forms of production organization and labor control in the United States and Europe (Aglietta 1979; Lipietz 1987) and imperatives of structural adjustment imposed on Third World states by a reconstituted international debt regime (McMichael 1992). The social and economic forces driving world capitalist restructuring are mirrored in the stagnation of formerly buoyant mass markets in advanced capitalist countries. Accordingly, many Third World states pursuing EOI in this global economic environment, "will find it increasingly difficult to maintain overall production and export levels" (Landsberg 1979, 61).

EXPORT-LED INDUSTRIALIZATION AND AGRARIAN CHANGE IN SINGAPORE, 1967–1985

The stars were aligned most favorably and
Singapore positioned itself accordingly.
—Former Prime Minister Lee Kuan Yew (February 1990)

Formerly a British colony and entrepôt integral to the shipment of regional foodstuffs and raw materials to Europe and the sale of European manufactures in Asia, Singapore became independent in 1965 after a short-lived merger with Malaya (Malaysia). Due to its colonial trading legacy, Singapore lacked a significant domestic industrial bourgeoisie with the requisite experience and capital to underwrite a postindependence industrialization campaign (Buchanan 1972; Rodan 1989). Limited further by a small domestic market (2 million in 1967), the lack of a hinterland given the break with Malaya, and the rise of economic nationalism in neighboring states, Singapore's economic growth hinged upon the ability of the People's Action Party (PAP) state to forge niches within the international division of labor and to develop new niches as world conditions changed. The late 1960s were precarious times in Singapore, characterized by high unemployment, burgeoning labor unrest, and widespread poverty—factors central to understanding the heavy-handed mode of social and economic regulation that arose to undergird EOI. By repressing labor unions, political opposition, and other individual freedoms, the PAP state created a social climate conducive to attracting foreign investment in Singaporean industry.

In short, the PAP state assumed a leading hand in ensuring the social, political and economic prerequisite for EOI. Moreover, it did this in such a way to significantly influence the allocation of factors of production. It contributed to Singapore's comparative advantage as an export base for low-skill, labour-intensive manufacturing production. (Rodan 1987, 155)

The period from 1967 to 1973 was one of exceptionally high economic growth, spurred by foreign investment in manufacturing, expanding world trade, the U.S. war effort in Vietnam, and the Indonesian oil boom (Buchanan 1972). The share of gross domestic product (GDP) from manufacturing rose from 17 percent to 24 percent, and real GDP increased by an average of 13 percent annually (King, Low, and Heng 1988, 7–8). The main impetus for manufacturing growth was external demand, with the fastest growing subsectors, such as electrical machinery and textiles, having a high export orientation. Singapore's tertiary sector was also expanding, with much of this growth linked to developments in the region. South Vietnam emerged as the fourth largest export market given Singapore's role in servicing the U.S. war effort (Buchanan 1972, 55), and the rise of the right-wing Suharto regime in Indonesia prompted a wave of investments by U.S., European, and Japanese firms in Indonesian oil exploration, with Singapore as the hub of their operations. Growth of the Singapore economy, therefore, was closely tied to the importance of the Malay world, and Southeast Asia in general, within the postwar U.S. political and economic calculus (Fifield 1973; Wu 1972).[1]

Uniquely, Singapore's industrial regime was *not* linked to the state sponsorship of a domestic farm sector. Agricultural policy centered on national food security and ensuring stable food supplies, but this was achieved in the absence of extensive state regulation and subsidization of agriculture or the rise of powerful agrarian regimes. Small, urban farms supplied most domestic food needs during the early years of EOI (Blaut 1953; Tempelman and Suykerbuyk 1983). Backyard or part-time farming was a key source of income for many Singaporeans, a legacy of the British prewar "grow your own food" campaign (Buchanan 1972, 82; Tempelman and Suykerbuyk 1983, 63). Local farms supplied all of Singapore's pork and egg consumption needs, from 75 to 80 percent of all chicken meat consumed, and about 25 percent of total vegetable consumption

Figure 10.1
Agricultural Self-Sufficiency Ratios, Singapore, Selected Products, 1972–1990

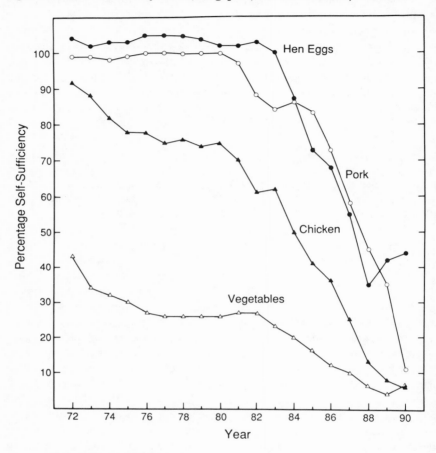

Source: Singapore Government
Primary Production Department.

(see Figure 10.1).[2] Residual food needs were met by imports and modest volumes of food aid.[3] Singapore was highly dependent upon imports to meet consumption needs in only a few commodities, notably rice. For this reason, in 1974, rice—a major wage food in Singapore—became the first agricultural commodity to be regulated under provisions of the Price Control Act.

Intensification of production during the 1970s allowed food self-sufficiency levels generally to remain high even as the amount of land allocated for agricultural purposes declined. There were 16,500 licensed farms and 12,000 hectares used as farmland in Singapore in 1973. This fell to 12,400 and 9,000, respectively, in 1979, as agricultural lands were rezoned for industrial and residential purposes (Neville 1992, 247). During the same time, per capita consumption levels of chicken, vegetables, and eggs rose appreciably (see Table 10.1). Self-sufficiency levels remained stable in pork and egg production but declined in chicken and vegetable production (see Figure 10.1).

During the early years of EOI, job creation was key, but as labor supplies tightened in the mid-1970s, emphasis was given to promoting capital-intensive industries (Rodan 1989). The focus on high-tech, capital-intensive industries reached full bloom during 1979–1985, the so-called Second Industrial Revolution. From 1973 to 1979, real GDP increased by an average of 7.3 percent annually, and manufacturing as a share of GDP continued to rise (Kng, Low, and Heng 1988, 6–8). Industrial growth was fueled by wage restraints and the greater incorporation of women and foreign workers in the labor force. In 1980 foreigners made up 11 percent of the total workforce (1988, 15), and the participation of women in the labor force rose from 30 percent in 1970 to 47 percent in 1980 (Singapore Department of Statistics 1983, 32).

Constraints on Singapore's industrial regime were reflected in concerns of maximizing domestic agricultural productivity to allow the movement of labor from agriculture to other sectors of the economy. Parallel arguments were made about informal food marketing, a "low-productivity and labour-intensive market sector [that] contributes little to the state's coffer" (Cheng 1990, 308). The informal food marketing subsector was increasingly viewed as incongruous with Singapore's changing position within the international division of labor (Liu 1991; Singapore Ministry of the Environment 1987). Food wholesalers, street hawkers, and small pig farms were cited as major polluters whose activities were contrary to "enhancing the confidence of our investors and attracting tourists to the country" (Singapore Ministry of the Environment 1987, 36).[4] From 1983 to 1986, over 4,900 street hawkers were licensed and resettled in food centers built by the Housing and Development Board, the Ministry of the Environment, and the Urban Redevelopment Authority. In 1984 fruit and vegetable wholesalers were resettled in a central market facility (1987, 26–27), and by the late 1970s, pig farming was limited to three peripheral areas of the island: Lim Chu Kang, Jalan Kayu, and Punggol. By the early 1980s, commercial pig production was restricted to large-scale, intensive pig farming estates in Jalan Kayu and Punggol (see Figure 10.2).

In 1979 the Second Industrial Revolution was initiated, the purpose of which was "to elevate Singapore's technological status in the hierarchy of the new international division of labor" (Rodan 1987, 150). It was believed that Singapore was losing its competitive advantage as a low-cost production site for labor-intensive manufactures, and that wage rates, which had been kept at artificially low levels to attract foreign capital, needed to rise in order to develop the skilled workforce that would attract new foreign investments in these industries.

By 1985, however, it was clear that the Second Industrial Revolution was not meeting expectations. Essentially, the PAP state miscalculated the degree to which Singapore could reposition itself within the currents of world capitalist restructuring (Rodan 1987, 1989). After twenty years of high growth, the Singa-

Table 10.1
National Consumption Levels, Selected Agricultural Products, Singapore, 1972–1990

Consumption

Year	Chicken		Pork		Vegetables		Eggs	
	Total (1,000MT)	Per Capita (kg/year)	Total (1,000MT)	Per Capita (kg/year)	Total (1,000MT)	Per Capita (kg/year)	Total (million)	Per Capita (# eggs/year)
1972	41.8	19.4	69.2	32.2	113.6	52.8	365.2	169.7
1973	42.1	19.2	64.2	29.3	108.6	49.5	397.0	181.0
1974	46.8	21.0	59.6	26.7	117.9	52.9	420.3	188.5
1975	49.9	22.1	59.7	26.4	122.6	54.2	450.0	198.9
1976	53.6	23.4	67.0	29.2	134.2	58.5	485.6	211.7
1977	59.2	25.5	69.4	29.8	129.1	55.5	492.4	211.8
1978	58.3	24.8	65.6	27.9	149.2	63.4	511.0	217.1
1979	60.8	25.5	69.4	29.1	138.8	58.2	503.3	211.2
1980	62.9	27.6	65.2	28.6	136.9	60.0	541.5	237.3
1981	70.4	30.3	61.2	26.3	153.2	65.9	522.3	224.7
1982	79.0	33.4	64.0	27.1	150.9	63.8	511.6	216.3
1983	77.5	32.2	67.7	28.1	153.1	63.6	513.6	210.2
1984	80.6	33.0	74.0	30.3	163.7	67.0	552.3	226.0
1985	78.2	31.5	75.0	30.2	181.4	73.1	557.2	224.4
1986	78.3	31.1	73.6	29.2	174.5	69.3	553.1	219.6
1987	84.4	33.1	76.6	30.0	168.8	66.1	586.7	229.7
1988	77.7	29.9	81.7	31.4	166.0	63.9	661.1	254.4
1989	76.6	28.9	81.2	30.7	167.4	63.2	768.0	290.1
1990	81.7	30.2	81.6	30.2	155.8	57.6	832.0	307.6

Source: Primary Production Department unpublished data; Singapore Department of Statistics (1983; 1991).

202

Figure 10.2
Former Pig Farming Estates and Agrotechnology Parks, Singapore

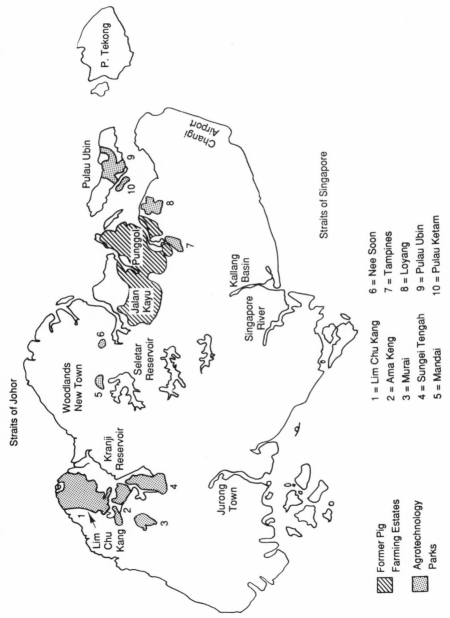

Straits of Johor

P. Tekong

Pulau Ubin

Woodlands
New Town

Kranji
Reservoir

Lim
Chu
Kang

Seletar
Reservoir

Jalan
Kayu

Punggol

Changi
Airport

Jurong
Town

Kallang
Basin

Singapore
River

Straits of Singapore

Former Pig
Farming Estates

Agrotechnology
Parks

1 = Lim Chu Kang
2 = Ama Keng
3 = Murai
4 = Sungei Tengah
5 = Mandai

6 = Nee Soon
7 = Tampines
8 = Loyang
9 = Pulau Ubin
10 = Pulau Ketam

pore economy was in a deep recession, and the PAP state faced a severe accumulation crisis as well as potent challenges to its political legitimacy (Rodan 1992). Singapore's merchandise exports began to flatten out in 1981 and declined in 1985 (Singapore Ministry of Trade and Industry 1986, 27–28). While real GDP rose an average of 8.5 percent annually from 1980 to 1984, it fell by 1.7 percent in 1985, and the unemployment rate rose to 5 percent, the highest level in fifteen years (1986, 38–39).

INDUSTRIAL RESTRUCTURING, 1985–1993

The severity of the 1985 recession prompted two government economic reports: the "Concept Plan" (Singapore Ministry of Trade and Industry 1986) and the "Strategic Economic Plan" (Singapore Ministry of Trade and Industry 1991) which, together, chart the course for the current restructuring of the national economy. Both reports place structural changes in external demand at the forefront of Singapore's economic woes. Global demand and capital-exporting structures changed during the 1980s, with direct foreign investment now being used to facilitate market access rather than as a means for firms via Third World subcontracting to remain competitive in home markets. Factories are increasingly established "where the goods will be sold . . . not in the countries where the goods can be most economically produced" (Singapore Ministry of Trade and Industry 1986, 11). Formerly buoyant mass markets in core countries have tightened, and international trade has stagnated, factors constraining the viability of Singapore's export-led industrial regime. Other external forces cited as affecting Singapore negatively include excess capacity in oil refining—"the prosperity and ebullience of the petroleum boom have ended" (1986, 5)—and low commodity prices affecting trade with other ASEAN countries.

The period of easy growth is now over. . . . Internally, we have finished doing all the easy things which can be done to foster growth. Externally, international trade is no longer expanding as exuberantly as it used to, and worse, the trading environment is becoming increasingly hostile. From now on, growth will be harder to achieve. (1986, 4)

The vision of the future is one of regained international competitiveness built upon a more flexible, multidimensional national economy (Singapore Ministry of Trade and Industry 1991). Attention is given to developing an array of high-value manufacturing and services industries, based upon both local and foreign capital, and enhancing tertiary infrastructures to provide skilled workers for these sectors. Manufacturing activities promoted include parts and components (believed to be less sensitive to protectionism) and high-value, low-volume performance-sensitive products. Services promoted include those relating to sea, land, and air transport, telecommunications, laboratory testing, and agrotechnology. Singapore is *not* becoming a post industrial city, but the nexus of manufacturing enterprises does seem to be changing, with linkages to a highly skilled, more fairly remunerated workforce. Economic restructuring, therefore, involves reorganizing "the way human and physical resources are managed" (Singapore Ministry of Trade and Industry 1991, 4).

Industrial restructuring is reflected in efforts to alter the urban and regional form to attract new international labor and capital flows and to spawn new forms of domestic consumption within the development of new social spaces. Specifically, Singapore is being redesigned to appear greener and more tropical- or island-like, characteristics perceived to be attractive to new international labor and capital flows—"it is necessary to maintain a high quality living environment to attract the right people to meet Singapore's needs" (Singapore Ministry of Trade and Industry 1991, 48). The Urban Redevelopment Authority's planning objectives for the year 2000 and beyond, articulated in the "Living the Next Lap" report, focus on the creation of greenery-laden industrial/services clusters framing a revitalized downtown-marina district. Emphasis is given to expanding domestic consumption of recreational and cultural activities, a notable shift away from the hegemonic sacrificial work ethic so central to the early EOI regime.

The Next Lap encourages us to think of a more gracious, balanced lifestyle. . . . Our common goal should be to create a city within an island which balances work and play, culture and commerce; a city of beauty, character and grace, with nature, waterbodies and urban development weaved [sic] together. (Singapore Urban Redevelopment Authority 1991, 6)

FROM PIGS TO ORCHIDS: AGRICULTURE IN THE NEXT LAP

The movement toward a "more gracious, balanced lifestyle" has far-reaching implications for the structure of agriculture in Singapore as well as for the matrix of Singapore's international food relations. Rising per capita incomes have led to a mass market for many fresh food products and to rising aggregate animal protein consumption levels (see Table 10.1). Even with efforts to intensify production, food self-sufficiency levels declined sharply during the 1980s (see Figure 10.1). Developing new social spaces for the consumption of leisure and cultural activities consonant with life in the next lap foster a further marginalization and physical compartmentalization of agriculture. Building nature reserves, greenways, and the like reduces the amount of land available for agricultural production, with accordant implications regarding domestic food security. Furthermore, externalities associated with traditional farming systems may impinge upon the quality of consumption within these new social spaces, thus generating pressures to refashion agriculture.

In 1983 the Ministry of National Development announced that domestic food self-sufficiency was no longer feasible given limited land and water resources, competing land use pressures, and the negative externalities associated with traditional urban farming systems. Factoring in all the costs, Singapore would be better off importing some foods rather than raising them locally.[5] The first action in this regard was the April 1985 announcement that domestic pig farming would be phased out (*Primary Production Bulletin* 1988a). This was followed by the approval of an Agrotechnology Park Master Plan in 1986 (Singapore Ministry of National Development 1989). According to the plan, farming in Singapore would be limited to 2,000 hectares designated for agrotechnology park development. Ten agrotechnology park complexes are slated for development, including one at Lim Chu Kang (a former pig-raising area) and others on Pulau Ubin, Pulau Ketam, and near Kranji Reservoir (see Figure 10.2).

The new agricultural policy reflects a dual strategy for meeting Singapore's fresh food needs. High-value, aesthetically pleasing and environmentally acceptable agricultural systems will remain in Singapore; other foods will be sourced via "agreements with various neighboring countries" (Cheng 1990, 209). Activities encouraged within the parks include aquaculture, intensive vegetable production, ornamental plant production, intensive layer (egg) systems, and research activities related to pig and cattle embryo transfer, artificial insemination, and plant biotechnology. Intensive vegetable and layer systems are designed to raise self-sufficiency levels in these products (see Figure 10.1 for early indications that this is being accomplished). The Primary Production Department (PPD) expects that these intensive production systems will be able to supply 15 percent of total chicken consumption and 87 percent and 20 percent of all eggs and vegetables consumed, respectively (Singapore Ministry of National Development 1989, 1). Aquacultural and horticultural systems are export oriented, and agrotechnology services are to be exported within tropical Asia, "as agricultural developments in the region gain impetus" (Singapore Ministry of Trade and Industry 1986, 188). In all cases, foreign capital is encouraged; although land use is based upon an open tender system, "foreign companies need not go through [this process] if they are involved in leading-edge agrotechnology" (Dove 1988).

Agricultural restructuring took place with great pace. By 1990, all pig farms were gone. Most farmers were resettled in new areas and new occupations, and several highly capitalized farmers were encouraged to raise high-value products like prawns and orchids (de Silva 1989). In September 1986, the first land tenders were offered in three of the parks (*Primary Production Bulletin* 1986a). By June 1993, farming activities were being carried out on 637 hectares, about one-third of all land designated for park development. In terms of net area, major farming activities include orchids and other ornamental plants, aquarium fish breeding, poultry for eggs, and vegetables/foodcrops (see Table 10.2). The ability of the PAP state to take such rapid measures to transform domestic agriculture—to ban pig farming, resettle farmers, and establish agrotechnology parks within a few years—illustrates the uniqueness of social and economic regulation in Singapore. There was no agrarian regime that could effectively resist state action.

Table 10.2
Farming Activities in Singapore Agrotechnology Parks, June 1993

Type of Farming Activity	Net Area (Hectares)
Aquatic Plants and Foliage	16.2
Aquarium Fish Breeding	104.9
Freshwater Food Fish	4.3
Marine Fish/Prawns	41.6
Aquarium Fish Exporter	6.0
Poultry (Eggs)	74.9
Ducks/Duck Breeding	39.0
Dairy Cattle	7.4
Goat Farming	2.0
Bird Breeding/Dog Breeding	37.7
Crocodile Farming	8.7
Holding of Imported Live Poultry	5.1
Vegetable/Foodcrops	53.7
Fruit Orchard	13.1
Mushrooms	4.0
Hydroponics/Nursery	19.2
Orchids/Ornamental Plants	163.5
Bean Sprout Production	3.3
Turf and Other	32.5
Total	637.2

Source: Singapore Primary Production Department, unpublished
        data.

SINGAPORE'S CHANGING FOOD IMPORT DEPENDENCE AND THE RECOMPOSITION OF REGIONAL PIG EXPORT SYSTEMS

The reverse image of greening in Singapore is the expansion of pig export systems in Indonesia, Malaysia, and Thailand. The banning of pig production has made Singapore entirely dependent upon imports to source the consumption of fresh pork, an important wage food for Singapore's ethnic Chinese majority. As domestic pig farming was being phased out, the PPD established stringent rules for the import of live pigs; only farms inspected and approved by the PPD may export pigs to Singapore. As of August 1992, 139 Malaysian farms, 13 Indonesian farms, and 17 Thai farms were approved by the PPD (Singapore Primary Production Department, unpublished data). Furthermore, in 1990, pork became subject to provisions of the Price Control Act, the second agricultural commodity to be regulated as such; pig importers must now be licensed (Chan 1990) and must sell a minimum number of pigs through the Hog Auction Market (HAM), a state-sponsored electronic auction system designed to control price fixing by large-volume importers (*Straits Times* 1990). These new regulatory structures link the eating patterns of affluent Singaporeans with Fordist pig export systems in the Southeast Asian periphery.

Table 10.3
Source of Singapore Pig Imports, 1988–1991

Year	Malaysia	Indonesia	Thailand	Total Import Volume
	(Percent	of total	imports)*	(# pigs)
1988	95	5	.2	555,939
1989	90	10	.1	697,807
1990	84	16	–	1,022,698
1991	78	22	–	1,242,207

* Percentage totals may not equal 100.00 due
to rounding error.

Source: Singapore Primary Production Department,
unpublished data.

In 1988 imports sourced about half of Singapore's fresh pork consumption
needs (see Figure 10.1 and Table 10.3). By 1990, the year after all remaining
Punggol farms were closed, Singapore was importing over one million pigs
annually (see Table 10.3). The source of imports was also changing, largely as
a result of state efforts to secure diverse sources of supplies to stabilize prices
and reduce dependence upon Malaysian farms, previously the main source of
imported pigs (*Primary Production Bulletin* 1988b; Singapore Ministry of National Development 1989, 1).[6] In 1988 Malaysian farms supplied 95 percent of
the Singapore market; Indonesian and Thai farms supplied 5 percent and 0.2
percent, respectively. By 1991 Malaysian farms supplied 78 percent of the market, and Indonesia's share had risen to 22 percent (see Table 10.3).

Commercial pig farms in Indonesia are located in three main areas—in the
Riau Islands, near Singapore; in North Sumatra, near Medan; and Near Pontianak, in Kalimantan (Borneo) (see Figure 10.3). Most Malaysian farms are located to the north and southeast of Kuala Lumpur. Commercial pig production
in North Sumatra and Malaysia predate the banning of pig farming in Singapore,
but the Riau Islands and Kalimantan are areas new to commercial swine production. Agricultural developments in the Riaus (which include the islands of
Batam and Bintan in addition to Bulan) must also be viewed within the context
of current efforts toward regional economic cooperation; the Riau-Singapore-
Johor (Malaysia) area is being promoted as the so-called Growth Triangle.

Two farms supply the majority of Indonesian pig exports to Singapore: PT
Sinar Culindo Perkasa, located on Pulau Bulan (Bulan Island), in the Riau Islands, and PT Astana Bajong Permai, located near Pontianak, in Kalimantan. In
1991, 94 percent of all pig exports from Indonesia came from these two farms
(see Table 10.4). The Pulau Bulan farm, which began as a joint venture between
the United Industrial Corporation (UIC) of Singapore and the Indonesian con-

Figure 10.3
Singapore and Surrounding Region

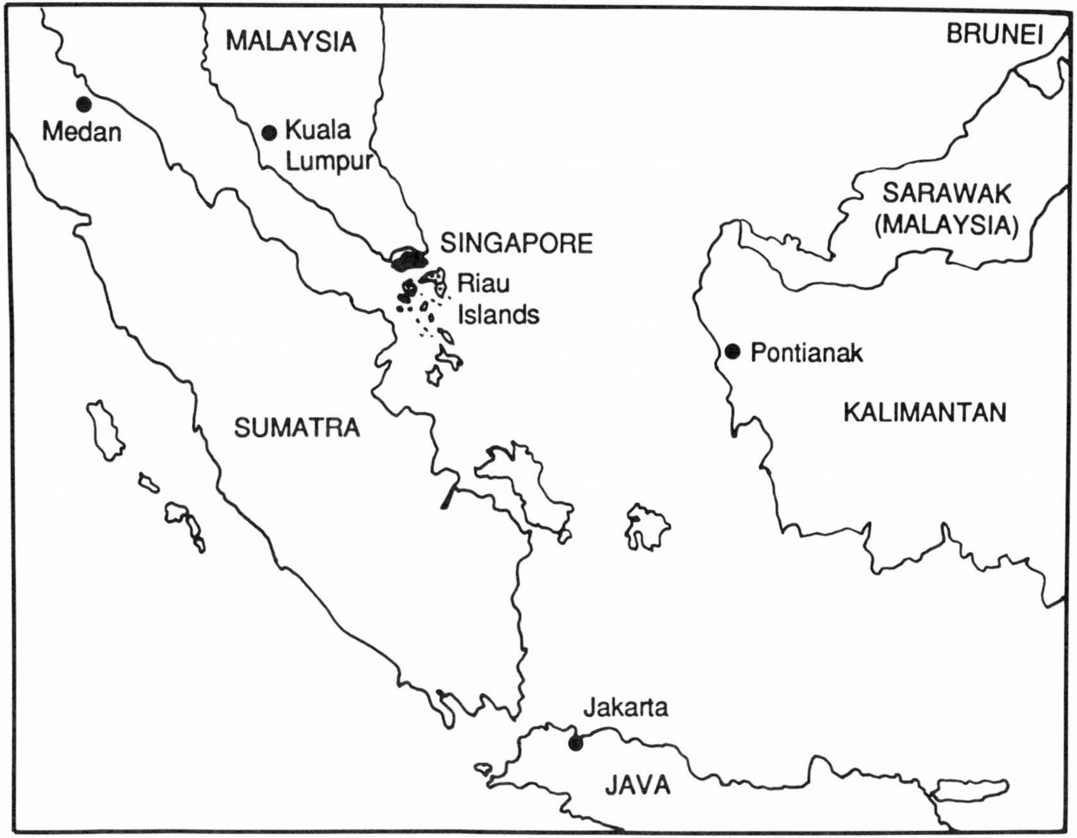

210 • Recomposition of Agro-Food Systems

Table 10.4
Source of Indonesian Pig Exports to Singapore, 1988–1991

Year	P.Bulan	Medan	Kalimantan	P. Bintan	Total Indo.
	(Percent of Total)*				Pig Exports
					(# pigs)
1988	91	9	-	-	24,716
1989	96	5	-	-	66,519
1990	66	11	23	.2	160,975
1991	61	5	33	.4	272,531

* Percentage totals may not equal 100.00 due to
  rounding error.

Source: Singapore Primary Production Department,
        unpublished data.

glomerates, the Salim Group and Sinar Mas, is the larger of the two.[7] It is part
of a new broad-based agro-industrial complex on the largely uninhabited 10,000-
hectare island leased by the Salim Group (*Singapore Business* 1990). Besides
pig farming, activities on the island include supplying broilers for regional Kentucky Fried Chicken outlets and the export of designer orchids (Schwarz 1990).
The Pulau Bulan farm, established in 1986 (*Singapore Business* 1990), uses
U.S. swine genetics (*Primary Production Bulletin* 1986b). The first load of pigs
was shipped to Singapore in August 1987 (*Primary Production Bulletin* 1987);
by 1991, the farm supplied 13 percent of Singapore's fresh pork consumption
needs (see Tables 10.3 and 10.4). Feedstuffs for the Pulau Bulan farm are sourced via Gold Coin, a Singaporean animal feed maker, in which the Salim Group
now has a 74 percent majority share (Schwarz 1990). The Kalimantan farm, a
single proprietorship, shipped the first load of pigs to Singapore in March 1990
(Nathan 1990), and by 1991, supplied 7 percent of the Singapore market.[8] The
15 million (U.S. $) operation includes an on-site feedmill and 4,000 hectares of
land devoted to corn production (Nathan 1990).

CONCLUSIONS

Efforts by the state to restructure the Singapore national economy in light of
accumulation crises resulted in a rapid and dramatic transformation of the social
structures of food production and consumption in the republic. These events, in
turn, have fostered a reconstitution of Singapore's international food relations.
In the wake of the 1985–1986 recession, the PAP state initiated a comprehensive
industrial and agricultural restructuring program to reintegrate the national economy within the contours of the changing world-economy. Until the mid-1980s,
Singapore's export regime was propelled by a heavy-handed mode of social and

economic regulation linking the interests of a corporatist state with the competitive crises facing industrial capital in core countries. The EOI regime was anchored in the repression of organized labor and the provision of cheap wage foods via domestic, urban farming systems supplemented by U.S. food aid and the regulation of rice imports. Post-1985 agricultural restructuring involved the phasing out of traditional farming systems and the limiting of domestic agricultural production to state agrotechnology parks. Furthermore, agricultural policy aims of securing high levels of domestic food self-sufficiency have been abandoned, and new regional agro-export supply zones have been established within the Southeast Asian periphery to source the fresh food consumption needs of Singapore's now affluent population. The expansion of pig export systems in Malaysia, Thailand, and Indonesia and the development of accordant regulatory structures in Singapore to monitor the pig trade and stabilize pork prices are illustrative of the nature of Singapore's international food relations within the next lap of economic growth.

The events in Singapore represent a particularly telling, mature case of the broader dynamics of contemporary agro-food system restructuring. Singapore is an unusually adaptive political economy because of the absence of powerful domestic agrarian regimes and its peculiar legacy of national social and economic regulation, factors central to understanding the rapidity by which agricultural restructuring programs were implemented during the 1980s and 1990s. As the former food regime, based upon U.S. hegemony in international agro-food transactions, wanes, the central elements constituting a new world food order are exposed. Grounded in the recomposition of markets, the hegemonic ideology of neoliberalism, and structures facilitating global accumulation by TNCs, the contemporary agro-food system is characterized by the proliferation of agro-export platforms in the new agricultural countries (NACs) to provision regional affluent markets with high-value foodstuffs, global sourcing by TNCs, and the erosion of institutions supportive of nationally based farm sectors (Friedmann 1993; McMichael 1992). The Singapore case sheds light on the role of the state in negotiating the globalization process by carving the terrain for the successful integration of the affluent Singapore market with regional agro-export systems, thereby presenting in bold relief the social and economic forces underlying contemporary agro-food system restructuring. The Singapore example may also provide a glimpse of the nature of future international agro-food transactions as the emergent food regime matures.

NOTES

1. U.S. influence on the Singapore economy follows that of Britain. During the 1960s, the British military presence on the island contributed a major share of GDP and employment. This waned when British troops were withdrawn in the early 1970s (Buchanan 1972, 86–89).

2. Pork, fish, chicken, and eggs are the main animal proteins consumed by Singaporeans. Pork is an important food for the ethnic Chinese majority; other animal proteins

consumed include duck, mutton, and beef. Local farms have supplied a major (but de-clining) share of total duck consumption, but beef and mutton must be imported. Rice and vegetables are also staples of the Singapore diet. This chapter focuses on the pro-duction and consumption of pork, chicken, vegetables, and eggs.

3. Prior to 1980, Singapore received Title II US PL 480 funds (the type of food aid given as grants). Amounts received (in US $ value) are listed below (Source: Janet E. Kodras, personal communication).

1959	$190,000	1966	$ 0	1973	$62,000
1960	206,000	1967	119,000	1974	46,000
1961	170,000	1968	178,000	1975	48,000
1962	225,000	1969	24,000	1976	59,000
1963	320,000	1970	154,000	1977	42,000
1964	0	1971	181,000	1978	0
1965	0	1972	51,000	1979	41,000
				1980+	0

4. Pig farms and food marketers were cited as major polluters within the 1977–1987 Clean Waters Campaign aimed at removing pollutants from the Singapore River and Kallang Basin (Singapore Ministry of the Environment 1987). These activities *were* pol-lutive, but the rhetoric of program planning and implementation confounds issues of water security and public health with issues relating to the development of new economic and social spaces for investors, visitors, and locals. The program, therefore, was linked to broader accumulation concerns.

5. These issues were elaborated to the author by the staff at the Ministry of the En-vironment and the Primary Production Department during fieldwork conducted in Singa-pore in June–July 1992.

6. This aspect was elaborated to the author by PPD Veterinary Regulatory and Agro-technology Services staff during interviews in Singapore in July 1992.

7. In a late 1990 takeover, the Salim Group, run by Indonesian magnate Liem Sioe Liong (also known by the name Sudono Salim), the so-called Cement King, purchased the assets of UIC (Friedland 1991).

8. Production is more concentrated on approved Indonesian farms than on those in Thailand or Malaysia. The PPD registers and monitors the production capacity of all approved farms (as measured by the number of pigs produced per month). As of August 1992, PPD records indicated that PT Sinar Culindo Perkasa and PT Astana Bajong Per-mai, with a combined production capacity of 20,000 pigs per month, accounted for 77 percent of the total production capacity of all approved Indonesian farms. The two largest farms in Thailand, with a combined production scale of 8,330 pigs per month, made up 25 percent of the total production capacity of approved Thai farms. The two largest Malaysian farms, with a combined production scale of 7,550 pigs per month, accounted for 7 percent of the total production capacity of all approved Malaysian farms. The largest twenty-five farms, with a combined production scale of 46,130 pigs per month, made up 43 percent of the total production capacity of all approved Malaysian farms (Singapore Primary Production Department, unpublished data). While production scale data are use-

ful in illustrating structural differences among Indonesian, Malaysian, and Thai farms, it is important to note that these percentages cannot be directly applied to measure the share of the Singapore pig market controlled by a farm or set of farms because these are not import-export data.

REFERENCES

Aglietta, Michel. 1979. *A Theory of Capitalist Regulation: The US Experience.* London: New Left Books.

Ariff, K. A. Mohamed, and Hal Hill. 1985. *Export-oriented Industrialization: The ASEAN Experience.* Sydney, Australia: Allen and Unwin.

Blaut, J. M. 1953. "The Economic Geography of a One-Acre Farm on Singapore Island: A Study in Applied Micro-geography." *Malayan Journal of Tropical Geography* 1: 37–48.

Buchanan, Iain. 1972. *Singapore in Southeast Asia: An Economic and Political Appraisal.* London: G. Bell and Sons Ltd.

Chan, Caroline. 1990. "Gov't to License All Pig Importers." *Straits Times*, May 26.

Cheng, Lim Keak. 1990. "Social Change and Fresh-Food Marketing in Singapore." *GeoJournal* 20,3: 301–10.

de Silva, Gerry. 1989. "From Pigs to Fish: High-Tech Fish Farm One of the Biggest in Region." *Straits Times*, February 17.

Dove, Richard. 1988. "Singapore Seeks Partners for High-Tech Farming Revolution." *Financial Times*, May 12.

Dulles, John Foster. 1950. *War or Peace.* New York: Macmillan.

Fifield, Russell H. 1973. *Americans in Southeast Asia: The Roots of Commitment.* New York: Thomas Y. Crowell.

Friedland, Jonathan. 1991. "Invisible Hand: Island Shopping." *Far Eastern Economic Review*, February 21, 45.

Friedmann, Harriet. 1982. "The Political Economy of Food: The Rise and Fall of the Postwar International Food Order." *American Journal of Sociology* 88: S248–S286.

———. 1991. "Changes in the International Division of Labor: Agri-Food Complexes and Export Agriculture." In *Towards a New Political Economy of Agriculture*, edited by W. H. Friedland, L. Busch, F. H. Buttell, and A. P. Rudy, 65–93. Boulder, Colo.: Westview Press.

———. 1993. "The Political Economy of Food: A Global Crisis." *New Left Review* 197 (January/February): 29–57.

Friedmann, Harriet, and Philip McMichael. 1989. "Agriculture and the State System: The Rise and Decline of National Agricultures, 1870 to the Present." *Sociologia Ruralis* 29,2: 93–117.

Hamilton, Clive. 1983. "Capitalist Industrialization in East Asia's Four Little Tigers." *Journal of Contemporary Asia* 13,1: 35–73.

Kng, Chng Meng, Linda Low, and Toh Mun Heng. 1988. *Industrial Restructuring in Singapore.* Asia Pacific Monograph no. 3. Singapore: National University of Singapore, Department of Economics and Statistics.

Kodras, Janet E. 1993. "Shifting Global Strategies of US Foreign Food Aid, 1955–90." *Political Geography* 12,3: 232–46.

Landsberg, Martin. 1979. "Export-led Industrialization in the Third World: Manufacturing Imperialism." *Review of Radical Political Economics* 2,4: 50–63.

Lim, Chong-Yah. 1983. "Singapore's Economic Development: Retrospect and Prospect." In *Singapore Development Policies and Trends*, edited by P.S.J. Chen, 89–104. Singapore: Oxford University Press.

Lipietz, Alain. 1987. *Mirages and Miracles: The Crisis of Global Fordism.* London: Verso.

Liu, Thai Ker. 1991. "Improving the Living Environment of Singapore." *Environmental Monitoring and Assessment* 19: 251–59.

McMichael, Philip. 1987. "Foundations of U.S./Japanese World-Economic Rivalry in the 'Pacific Rim'." *Journal of Developing Societies* 3,1: 62–77.

———. 1992. "Tensions between National and International Control of the World Food Order: Contours of a New Food Regime." *Sociological Perspectives* 35,2: 343–65.

McMichael, Philip, and David Myhre. 1991. "Global Regulation vs. the Nation-State: Agro-Food Systems and the New Politics of Capital." *Capital and Class* 43: 83–105.

Nathan, Francesca. 1990. "First Shipment of 1,200 Pigs Sail in from Kalimantan." *Straits Times*, March 10.

Neville, Warwick. 1992. "Agribusiness in Singapore: A Capital-intensive Service." *Journal of Rural Studies* 8,3: 241–55.

Primary Production Bulletin. 1986a. "PPD Launches Plans to Develop Agrotechnology Parks." September, 1.

———. 1986b. "Breeder Swine Transshipped thru S'pore." October, 13.

———. 1987. "First Shipment of Pigs from Pulau Bulan Arrives." September, 3.

———. 1988a. "Phasing-out of Pig Farming." April, 1.

———. 1988b. "Mission to Source Hog Supply from Indonesia." September, 20.

Rodan, Garry. 1987. "The Rise and Fall of Singapore's 'Second Industrial Revolution'." In *Southeast Asia in the 1980s: The Politics of Economic Crisis*, edited by R. Robison, K. Hewison, and R. Higgott, 149–76. Sydney, Australia: Allen and Unwin.

———. 1989. *The Political Economy of Singapore's Industrialization: National-State and International Capital.* London: Macmillan.

———. 1992. "Singapore Leadership Transition: Erosion or Refinement of Authoritarian Rule?" *Bulletin of Concerned Asian Scholars* 24,1: 3–17.

Schwarz, Adam. 1990. "Indonesian Conglomerate Heads for Free Trade Zone: Export Platforms." *Far Eastern Economic Review*, October 18, 77–88.

Singapore Business. 1990. "Indon Group to the Fore." December, 50–52.

Singapore Department of Statistics. 1983. *Economic and Social Statistics: Singapore 1960–1982.* Singapore: National Printers Ltd.

———. 1991. *Yearbook of Statistics.* Singapore: National Printers Ltd.

Singapore Ministry of National Development. 1989. "Master Plan for the Agrotech Industry." *Productivity Digest*, January, 1–4.

Singapore Ministry of the Environment. 1987. *Clean Rivers: The Cleaning Up of Singapore River and Kallang Basin.* Singapore: Stamford Press Pte. Ltd.

Singapore Ministry of Trade and Industry. 1986. *The Singapore Economy: New Directions.* Singapore: National Printers Ltd.

————. 1991. *The Strategic Economic Plan: Towards a Developed Nation.* Singapore: National Printers Ltd.

Singapore Urban Redevelopment Authority. 1991. *Living the Next Lap: Towards a Tropical City of Excellence.* Singapore: Shing Lee Publishers Pte. Ltd.

Straits Times. 1990. "Pork Prices Will Be Made Available Weekly." May 30.

Tempelman, G. J., and F.J.J. Suykerbuyk. 1983. "Agriculture in Singapore: Problems of Space and Productivity." *Singapore Journal of Tropical Geography* 4,1: 62–72.

Wu, Yuan-li. 1972. *Strategic Significance of Singapore: A Study in Balance of Power.* Washington, D.C.: American Enterprise Institute.

11

Gender and Cheap Labor in Agriculture

Jane L. Collins

The 1980s and 1990s have seen many changes in agricultural labor relations throughout the world. Agribusiness firms are increasingly employing female workers for the jobs previously performed by males. They are relying, more than in the past, on "undocumented" workers.[1] At the same time, for some crops, they are moving toward production subcontracting—a practice that relies on "family labor." These changes have been implemented in order to lower the cost of wages, to overcome technical constraints, to transfer production risks, and to enhance control over labor. They have had profound implications for the political struggles of agricultural workers and for the social reproduction of laboring classes in agriculture.

This chapter focuses on the growing employment of women by agribusinesses, particularly in the production of fruits and vegetables and other crops that require careful handling and timely performance of production tasks. I will make a series of arguments about this trend and its implications. First under consideration are the specific kinds of jobs for which firms recruit women and the kinds of contexts in which they employ them. In those particular crops where women are predominantly employed, firms seek high-quality labor at low cost. Ensuring that workers perform tasks in a careful and timely manner requires the exercise of extensive discipline and control. At the same time, minimizing wage payments requires the erosion of worker rights. By employing women, firms are able to obtain both of these goals simultaneously. Agribusinesses use gender ideologies to erode stable employment and worker rights where women are concerned. Of equal significance, employing women provides the employer with a way of invoking institutions beyond the workplace to extend and reinforce labor discipline.

This argument subverts attempts to speak simply about the impacts of women's (often) new-found employment in agriculture. It is clear that, in many cases, women's income rises significantly as a result of their employment, and that their income takes on greater importance within the household, sometimes enhancing their autonomy and power within the family. Studies documenting the different spending patterns of men and women indicate that women's income is more often spent on food and other basic necessities, and thus it has more direct effects on the health and well-being of children and other dependent household members (Dewalt and Dewalt 1992). Studies also indicate that the significance of women's income is increased by more general patterns of economic crisis, unemployment, and declining real wages that have meant that the male wage, in multiple-adult households, is often reduced or absent (McAfee 1991; Sen and Grown 1987; Aguiar 1990). While all of these things may be true and of great importance, it is precisely women's central role in providing for families and their responsibilities for children's health and well-being (particularly in times of economic difficulty and declining male wages) that make them vulnerable to firm strategies to degrade the conditions under which they labor. The expectation that women will do what is necessary to socially reproduce the household mitigates against their ability to struggle in the workplace, and it can have the effect of socializing them to the demands of employers.

Case studies of women's employment in agriculture have begun to demonstrate that as women work through and resolve these contradictions—between a "family morality" (Stolcke 1984) that defines their role as in the home, on the one hand, and their growing responsibilities outside the home, on the other; between the family's growing dependence on women's wages for basic needs and the firm's pressure on the level and regularity of their wages; between the devaluation of their work and its importance to firms—the conditions that make their labor cheap and malleable are eroded. In many contexts, women have asserted their right to make decisions at home and to exercise greater control over their own time and labor. Additionally, as pressures on women increase and their self-confidence in the workplace grows, they have begun to contest the conditions under which they work. The outcomes of these struggles to date have been diverse. Women become empowered politically and sometimes they gain rights. At the same time, in doing so, they negate the very characteristics that have made them attractive to firms—a move that is fraught with danger when the firms involved are transnationals with diverse holdings, who can (in some cases) transfer operations to other regions where labor is still cheap and docile.

The changes in agricultural labor processes to be described have emerged in a new competitive environment in which agribusiness corporations, like their industrial counterparts, have sought to "rationalize" their labor force by increasing its flexibility. Labor in agriculture has always been "casual," or seasonal and intermittent, but the evidence is that it is becoming more so (Pugliese 1991). Some researchers have suggested that, within agriculture as within industry, recent economic trends have led to a more homogeneous workforce. Much of the literature on economic restructuring has focused on the erosion of distinctions between "primary" (privileged) and "secondary" labor markets in contexts of casualization and subcontracting of work.[2] Whether the labor relations associated with restructuring are seen as providing new opportunities, or as undermining the security of the working classes, the assumption is that they are linked to the erasure of differences among workers, with each individual becoming a separable and replaceable component of a constantly adjusting system of production. Descriptive accounts of flexible production make it clear that it is most often women who are doing industrial homework, and women and immigrants who are most frequently employed on a casual or intermittent basis by industries. To date, however, these observations have not been incorporated into theoretical accounts of the forms flexible specialization takes, the places it operates, or its transformations over time. The evidence presented here suggests that, while branches of agriculture are experiencing pressures to increase the flexibility of their labor supply, such pressures do not demand homogeneous workers. Rather, flexibility for employers *depends* on the manipulation of specific features of workers.

Flexible labor practices in agriculture, as elsewhere, require a disenfranchisement of workers. As Truelove (1989) has pointed out, much of the literature on flexible specialization has emphasized the opportunities provided for entrepreneurs and has ignored the fact that flexibility is obtained by eroding the rights of workers to stable, full-time, year-round employment. This erosion requires an enhancement of control over workers—new sets of ideologies and practices that can effectively manage short-term or intermittent employees and that can deploy their labor to tasks that, although seasonal, often require skill, care, and sustained exertion. As corporate eyes scan the global landscape, seeking the best fit between production imperatives and labor force characteristics, gendered social relations provide a crucial tool for segmenting populations whose members are equally willing and able to perform tasks at hand. They also, as previously suggested, offer the prospective employer a set of institutions beyond the workplace that can extend and reinforce labor discipline.

It has long been understood that capitalist social relations are not reproduced solely through the circuits of capital and through relations in the workplace. Feminist and critical theorists have called attention to the ways in which a broad array of social, political, and market institutions sustain (and contest) the process of capital accumulation. Families, community forms, religious institutions, neighborhoods, cultural practices, the welfare state, laws and regulations—all have been analyzed from this perspective. The literature thus produced has suffered from two major shortcomings. On one hand, much of the early work in this area—that produced by Marxist feminists for example—tended to emphasize the functionality of diverse social forms to capital or capitalist classes. Western patriarchy was seen as useful to, but somewhat separate from, the projects of capital. Family forms were seen as important, but still analyzed mechanistically—as adjunct or external. On the other hand, research that has emphasized the more integral relation of social institutions to capital—for example, French regulation theory or theories of the "social structures of accumulation,"—has tended to emphasize the role of the state and its agencies, the rules and agreements governing financial transactions, and policy, without giving much attention to the gendered nature of social relations or to family forms.

Several recent works have broken new ground in this regard. Mingione (1991), for example, has produced a highly detailed examination of the ways in which wages, as well as the demand for particular kinds and amounts of labor, link patterns of social reproduction to cycles of capital under particular industrial regimes. He presents a framework within which patterns of social reproduction and of capital accumulation in a given historical period are mutually constitutive. Mingione emphasizes that, if wages and labor market conditions broadly determine what is possible in the arena of social reproduction, the strategies of social reproduction also broadly determine the cost, quality, and quantity of labor that will be available in the present and future generations, and they determine the capacity for consumption of the commodities produced by capital (1991, 128).

Wage rates and the availability of employment are clearly important determinants of the survival strategies of families. They present the additional advantage of being susceptible to investigation using economic surveys and indicators. There are still other ways in which family life and work life (or patterns of reproduction and cycles of capital accumulation) are integrally related, however. Gershuny (1983), Pahl (1985), and Wheelock (1990), for example, have studied the ways in which household members renegotiate their use of time and their consumption practices with the decline of formal sector employment. These accounts suggest that gendered social relations structure interactions, not just within the home, but within the workplace. Work routines performed for agribusiness, and those associated with the social reproduction of the family, impinge on one another both spatially and temporally. Conversations about gender and family—between male managers and female workers, and among men and women in the home—construct shared understandings and contested arenas that crosscut the work/home divide. Work and home are not separate spheres requiring different theoretical specifications, and they are not simply "mediated" by wages and labor markets. They are temporally and (sometimes) spatially differentiated sites within a single (conflict-ridden) system of social reproduction.

THE FEMINIZATION OF THE
AGRICULTURAL WORKFORCE

Shifts to a largely female workforce have not characterized all branches of agriculture. Firms in different sectors have adopted labor arrangements in response to the distinct sets of competitive pressures that emerge in producing different commodities and in response to the technologies available for producing various crops. In wheat, oils, livestock, and feed grains, for example, recent trends have favored labor displacement. For other commodities, including fresh and processed fruits and vegetables and some traditional export perennials, firms continue to use labor intensively and to experiment with new ways of managing it. In some cases, the technology to preserve the quality of delicate produce is not available; in others, the cost accounting suggests that where labor is cheaply available, labor-intensive production methods remain more lucrative.

Theories of comparative advantage in agriculture have always held that tropical countries should produce fruits and vegetables. While the original advantage possessed by countries at tropical latitudes was held to be climate, abundant and cheap labor soon became equally important to certain branches of agriculture. The U.S. agricultural doctrine that the production of labor-intensive crops, such as fruits and vegetables, should be left to the developing world was perhaps first articulated by the Nixon administration's Commission on International Trade and Investment Policy in 1971, but it is echoed by more recent documents from the U.S. Department of Agriculture as well:

production of fruits and vegetables is still labor intensive and furthermore, processing technology is mature and widely available. Thus, developing countries with relatively cheap labor will find it increasingly attractive to produce large quantities of fruits and vegetables for export to developed countries in both fresh and processed forms. (Sarris 1984, 4).

In addition to fruits and vegetables, certain other labor-intensive crops are seen by theories of comparative advantage as particularly appropriate for production in the developing world. These include some traditional export crops (tea, tobacco, and coffee, for example) and, more recently, hybrid seed production.[3]

Consumption of fruits and vegetables has increased steadily in the United States since World War II (Sarris 1984, 6). U.S. imports more than doubled between 1979 and 1989, from a dollar value of just over 2 billion to 4.8 billion.[4] During the same period, the value of Mexico's exports of fruit and vegetables increased from 500 million to 837 million; Brazil's from 457 million to 1.2 billion; and Chile's from 184 million to 676 million (Food and Agriculture Organization 1985, 1989).[5] In addition to consumption increases in the industrialized countries, there has been a growth in consumer demand in the developing world, often within producer countries themselves.[6] In the latter case, there has been not only a quantitative increase in demand, but also a qualitative shift in the kinds of products purchased. In many of the newly industrializing countries, the growth of the internal market has been linked to the expansion of supermarket chains at the expense of local (street) vendors, to the education of consumers to expect the year-round availability of products formerly available on a seasonal basis, and to expectations of goods with a uniform, unblemished appearance.

As fruit and vegetable firms have expanded production, they have shifted to a largely female labor force in many parts of the world. It is difficult to find large-scale surveys of labor force participation in agriculture that are broken down both by crop and by gender of workers. The evidence from case studies, however, overwhelmingly supports the prevalence of such shifts. A study of Del Monte's vegetable production in Mexico's Bajio Valley indicated that 75 percent of the workers employed directly by the company were women (Burbach and Flynn 1980, 188). Barrón (1991), reporting on a survey conducted in six states in Mexico,[7] found that over half of the labor force in fruit and vegetable production was made up of women.[8] Roldan (1982) reports that women make up from 90 to 95 percent of the labor force in packing, and from 50 to 60 percent of the harvest labor force in vegetable production in the state of Sinaloa, Mexico. Arizpe and Aranda (1981) have noted the predominance of women in the Mexican strawberry industry, and Ronner and Muñoz (1978), in Mexican grape production. In Chile, where the fruit industry contributes almost 20 percent of national exports, the labor force is predominantly (and increasingly) female (Lago 1987; Sáez 1986, 34; Lago and Olavária 1982; Goldfrank 1991, 3; Gomez 1991, 17). Women are an important component of the labor force in fruit and vegetable production in the Dominican Republic (Mones and Grant 1987; Raynolds 1991, 18), in Brazil (Collins 1993), in Senegal (Mackintosh 1989), and in the export flower industry (Gruhn 1991, 20; Medrano 1980; Silva and Corredor 1980).

The reason for fruit and vegetable firms turning increasingly to women workers appears to lie in the complex intersection of several production imperatives which they see as central to the success of their operations (Collins 1993). These imperatives are quality, flexibility, and political stability. For reasons discussed below, the employment of women provides firms with the opportunity to achieve goals simultaneously in these three areas.

The requirement for high-quality labor (a term coined by Wells, 1984, in her discussion of California strawberry production) refers to the need to handle products in a way that will allow them to meet demanding grade and quality standards for freshness, fragrance, appearance, color, weight, moisture content, shape, odor, and absence of blemish.[9] Ensuring that produce meets these standards requires high levels of labor supervision, or some other mechanism, to ensure that workers monitor the health and growth of plants, handle fruit carefully, and work quickly and efficiently. Agribusiness managers frequently argue (like their industrial counterparts) that women are both more "nimble and dexterous" and more "docile," meaning that they can be trained and counted upon to handle produce with care and attention. It is also believed that women are less likely to chafe under the heavy supervision required to ensure consistent levels of quality.[10]

The second imperative—flexibility—refers to the ability of firms to tailor labor supply to production cycles, hiring workers on a casual or intermittent basis, without long-term contracts, and without commitments to provide benefits, to respect seniority, or, in some cases, even to adhere to protective legislation. Agribusiness firms, like many of their industrial and service sector counterparts, have found it easier to justify (and negotiate) flexible strategies with women employees. Firm managers argue that seasonal or intermittent employment practices fit well with women's primary responsibilities to home and family.[11] Flexibility is always, of course, from the point of view of the firm's needs. It may entail working twenty-hour days during the harvest—a practice hardly consistent with the care of home and children.

The third and final imperative is political stability. In conjunction with the first two imperatives, firms seek ways to avoid the risks of the land invasions, strikes, and slowdowns that have historically occurred in situations where they exercise direct ownership and employ wage labor. They believe that hiring women is a good way to achieve this goal, and they are guided, in this assessment, by their observation that women are less likely to be members of rural unions, and less likely to take leadership or activist roles when they do join.

Lago and Olavária summarize many of these arguments as they are used by Chilean fruit firms: women are less likely to damage the fruit; they are "more responsible; they don't miss work; they are more submissive when reproached; they will accept any salary, and type of work and under whatever conditions" (1982, 189, my translation).

The employment of women helps agribusiness firms *combine* flexible employment strategies and a politically stable workforce, on the one hand, with their need for a high quality of labor, on the other—imperatives that have traditionally been at odds with one another. Generally, in the history of labor relations, quality labor has been obtained by employing workers for long periods of time, investing in training and the development of skills, and providing a graduated system of rewards as an incentive to good work. Such a strategy is obviously at odds with flexible employment practices. It also provides workers with continuity of contact (prerequisite to the development of a sense of community and to political action) and with skills that can serve as bargaining points—thus undermining political stability. The employment of women overcomes the contradictions among these three production imperatives.

Why or how do women seem to offer this combination of features? The answer seems to lie in the fact that employing women provides the firm with a way of invoking institutions beyond the workplace to extend and reinforce labor discipline. Just as Thomas (1985) has argued that lettuce companies employing undocumented workers in southern California are able to mobilize nonmarket forms of control in order to discipline workers (via the state, in the form of the Immigration and Naturalization Service), fruit and vegetable firms who employ women draw benefit from nonmarket forms of control located in the family. This process has been referred to by Stolcke (1984), in the southern Brazilian context, as "the exploitation of family morality."

CONTROLLING WOMEN WORKERS

There is an inherent difficulty in writing in general terms about the kinds of control exercised by firms over women workers and about the responses of the workers themselves.

[W]hat is problematic about this kind of use of "women" as a group, as a stable category of analysis, is that it assumes an ahistorical, universal unity between women based on a generalized notion of their subordination. Instead of analytically *demonstrating* the production of women as socioeconomic political groups within particular local contexts, the analytical move limits the definition of the female subject to gender identity, completely bypassing social class and ethnic identity. (Mohanty 1991, 64)

In the situations we are concerned with, where agribusiness firms employ new pools of female labor, women's experiences are shaped by many factors other than gender, and their lives (and the implications of new wage work for their lives) vary widely. The reason it makes sense to ask the question "Why does agribusiness recruit women?" is not because women are a homogeneous group who will experience the same effects, but because the fruit and vegetable firms under discussion *view* women as a homogeneous group whose employment can provide certain advantages. Firms have their own versions of family morality built up out of the national culture and class experience of management personnel, and the preconceptions are encoded in standard operating procedures and policies (Enloe 1989). Their encounter with women whose home responsibilities, workforce participation, and political lives differ from what they imagine in their corporate boardrooms generates many contradictions and resistances.

Agribusiness firms rely on gendered social relations to discipline and control women in three distinct kinds of ways. The first is technical control over the production operation—the ability to determine the ways in which women will perform tasks and the speed with which they will do so. The second is control over worker behavior—the regularity of attendance at work, promptness, willingness to work overtime when needed, and compliance with rules. The third aspect of the docility firms seek is the absence of contestatory behavior—running the gamut from complaints and confrontations, to worker organization, and actions against the firm. Each of these aspects of worker discipline is secured in different ways by gender ideologies and women's responsibilities in the home.

The literature reveals that many of the factors that are said to render women more compliant as employees, and that limit their political activity, are related to the fact that women retain primary responsibility for caring for children and the sick and elderly when they undertake waged work. Many authors argue forcefully that the difficulty of combining such responsibilities mitigates against women's participation in labor unions or other workplace organizations. Spindel (1987) says of women agricultural workers in southern Brazil that their "participation in union activities is marginal. Their explanation for this . . . is that they simply do not have the time" (1987, 64). For rural Nicaraguan women, Padilla, Murguialday, and Criquillón (1987) argue that "the fourteen to sixteen hour double day of domestic and wage work leaves women little time to participate in organization activities or even to rest" (1987, 135).

A more complex argument relates women's continuing family responsibilities to their view of themselves as workers. This argument, which has been most thoroughly developed for contexts in Latin America, points to the high value attached to women's work as mothers (and daughters) and the degraded status attached to rural wage work. Because of this valuation, women who work in agriculture may be reluctant to see themselves as "workers" in the same way that men do. They may continue to see their work as subsidiary and supplementary even when it accounts for a large proportion of family income.

Where women view their work as subsidiary, it is argued, they will not feel empowered to struggle for improved conditions. Whatever a woman brings in the way of income is "extra"—it need not be a fair wage or a family wage. As Kessler-Harris (1990) has argued so powerfully for women's work in the United States, when cultural precepts tell us that women work as mothers and daughters in order to "help," the emphasis is not on whether remuneration is just, but on what the wage can buy—its immediate impact on the family's standard of living. This justifies women's lower wages—a man's wage is based on what the market will bear, but a woman's wage is based on her need. These views structure a woman's sense of her rights as a worker, and what she feels it is possible and necessary to ask for. Additionally, within such a frame, a woman's wages and the conditions of her work do not reflect her worth as a person—that worth is derived from the health and well-being of her family. Compliance at work is not humiliating, as it might be for men, but a sacrifice that can produce a sense of accomplishment as it redounds to the benefit of the family.

An additional way in which it has been argued that gender inequality in the broader cultural context can generate compliance in the workplace is related to the different ways in which practices of labor control are understood when they are exercised on men and women. In agriculture, it is generally the case that managers continue to use what Edwards (1979) and others have called "simple control"—that is, foremen and managers "exhort workers, bully and threaten them, reward good performance, hire and fire on the spot, favor loyal workers and generally act as despots" (1979, 19). Thomas (1985) suggests that these practices of "simple" control are experienced differently by men and women workers. He describes situations in which foremen attempted to intimidate women workers through displays of anger or physical force, accompanied by threats of firing. Women responded to these incidents with embarrassment, rather than hostility; male workers noted, "To a man, that would be a challenge to fight" (1985, 180).

The fact that women continue to take primary responsibility for the care of family members in the home, and perhaps to see their work as subsidiary, should not overshadow the economic significance of their wages in many contexts. While the literature frequently notes that women's wages in agriculture may be central to family survival, it has not generally made a *connection* between that centrality and women's "compliance."

In contexts where women's labor *in* the home is highly valued, women's willingness to work for wages outside the home has been shown to be related to periods of economic difficulty. Several studies have shown that women's labor force participation increases with economic crisis in the Latin American context (Oliveira 1990; Prates 1990; Spindel 1990). These studies have argued that women's search for employment is related to male unemployment or to the reduced purchasing power of the wage under hyperinflation or the removal of price controls. This more general point has also been made quite specifically in the context of women's employment in Chilean fruit production.

[T]he pattern of female employment is a function of both the labor requirements of the fruit production cycle and, maybe more important, the high level of male unemployment . . . a consequence of neo-liberal model policies in agriculture. To the extent that male earnings are eroded and men have fewer possibilities of permanent employment, women are forced to accept non-traditional jobs . . . This phenomenon fits well with the specific labor needs of the fruit industry. (Lago 1987, 27)

This passage is especially interesting in noting that Chilean women are being employed in nontraditional (read formerly male) jobs while their husbands remain unemployed, and that this accords with the "specific labor needs" of the fruit industry. In another passage, the author speaks at greater length about the effects of this situation: "[B]ecause women are acutely aware of how important their contribution is to household income (because they have greater possibilities of employment than do men in this region), they accept exploitative conditions" (Lago 1987, 30). The same author quotes a fruit worker as saying: "[H]ere one doesn't speak of making a complaint because you are afraid that in the next year they won't receive you, and the money is needed for the home, to eat and dress yourself" (Lago 1982, 191, my translation).

While often related to the economic downturn, structural adjustment, and male unemployment, the need that propels women into waged work in agriculture may stem from other factors as well. In the research I conducted in a newly irrigated region of northeastern Brazil, it was not economic crisis, but a process of regional economic development that had jeopardized the income of some families and made women's work central. Fruit and vegetable production was introduced to the São Francisco Valley with irrigation in the 1970s. At that time, a massive dam was built at Sobradinho, flooding over 4,000 square kilometers and displacing 64,000 people. Although some attempts were made to resettle displaced families on government-funded irrigation projects, most people had to leave their homes before the projects were functioning. Thus, the majority of people whose lands were flooded either migrated to southern cities or colonization zones, such as Rondonia, or settled in nearby squatter settlements (Duque 1984; Andrade 1984). When newly irrigated lands were opened for corporate investment, firms hired women from these families to work in grape and mango production. This is not an uncommon pattern. Lago reports that many of the women employed in the fruit industry in Chile come from households "that have either lost their land or never owned land" (1987, 28). Mackintosh (1989) indicates that many individuals (both male and female) who worked for Bud-Senegal had lost the land they had previously cultivated to the corporation when it took up operations.

When women are working as a last-resort effort to meet the needs of the household, they are extremely vulnerable as employees. Their willingness to work long hours during some parts of the agricultural cycle and to be effectively laid off in others, to be hired on short-term contracts that are not covered by protective legislation and do not provide benefits, to receive lower wages than men, and to put up with intrusive forms of monitoring and control of their work are related to their need to mitigate the loss of a male income or the declining purchasing power of the male wage. Economic vulnerability, combined with an emphasis on work as a way of meeting responsibilities for the well-being of family members, thus disciplines women who work in agriculture and redounds to the benefit of the firm.

The lack of availability of alternative employment opportunities may exacerbate this situation. Thomas (1985) has described how women's relative lack of employment alternatives has made it possible for lettuce firms to recruit them for the lowest paid jobs in the lettuce industry. Many of the case studies of women agricultural workers cited above emphasize that these women have never worked outside of the agricultural labor market (Barrón 1991, 11; Arizpe and Aranda 1981, 460). Such a lack of alternatives may not always be an issue for women, particularly in areas with a thriving informal sector; however, it is a powerful factor shaping women's compliance in many rural areas.

CONCLUSIONS

As agro-food corporations have expanded their production of fruits and vegetables over the past two decades, they have recruited a largely female workforce for field work and packing. This is in part because women's labor in many parts of the world is still the "cheapest of the cheap," not only because women can be paid lower wages but because women's labor can be obtained on more flexible terms than men's labor in many cases. As we have also seen, firms employ women because they can more effectively control them in the ways necessary to meet the quality standards that apply to fruits and vegetables destined for export, and because they are seen as less politically volatile than male workers. It is because firms perceive women workers to be offering them the *combination* of quality, flexibility, and political stability that women are preferred in fruit and vegetable production.

As Enloe (1989) has said, however, labor is never simply "cheap"—it is made cheap, and the same goes for flexibility and political stability. Firms rely on family and community institutions to make women's labor cheap and to do part of the work of disciplining women workers. They rely on what Stolcke (1984) has called "family moralities" to teach women that their waged work is subsidiary to their work in the home; it is less important than men's work, and it is always in service to the requirements of social reproduction.

There is, however, a gaping contradiction between these family moralities and the real-life conditions experienced by many of the women working in agriculture. For the women working in fruit production in Chile, whose husbands and partners are unemployed, or the women heads of household working in vegetable production in northern Mexico, a notion of their work as subsidiary or secondary no longer makes sense. While a view of women's work as "helping" may be hegemonic in many of the contexts where agribusiness is expanding, it is not uncontested, and in fact, it may be highly unstable. Women see where their income goes, and when it is spent on food or fuel, or even school supplies, they may come to resent its definition as nonessential—even when it remains a small proportion of the family income. Where women become primary wage earners, when inflation renders a male wage increasingly inadequate, and when "temporary" labor becomes a permanent feature of a woman's life, statements about the subsidiary nature of women's work can only be tinged with irony or nostalgia. Most important, when women view their most important role as caring for their family, and when the conditions of their labor make that impossible, they bring to their workplace struggles, not just notions of fairness or justice, but their deepest concerns about family survival. This has been the case when women fruit, vegetable, and flower workers have, in fact, organized to demand wage increases, child care, or benefits in Chile, southern Brazil, Colombia, and Senegal.

NOTES

1. Undocumented workers include illegal immigrants, as well as individuals recruited for work outside the framework of government regulation (that is, without being officially registered as an employee and thus subject to labor codes, social security, and so on).

2. See Thomas (1985, chaps. 1 and 2) for a critique of this view.

3. Rice (1991, 10) has described the process whereby U.S. and European seed firms shifted production of hybrid seeds to Japan in the period before 1960, then to Taiwan when labor became costly and unavailable in Japan, and most recently to "inexpensive labor markets" like Mexico, Guatemala, Costa Rica, Chile, Thailand, the Philippines, and Indonesia.

4. Values are given in 1982 dollars.

5. It is important to note that these figures are in dollars and that growth in volume of trade went hand in hand with "substantial deterioration in export prices" during this period (Sarris 1984, 43).

6. World exports of fresh fruit grew at an annual rate of 9.6 percent from 1965 to 1975 and 7.7 percent from 1975 to 1985. World trade in fresh vegetables grew at average annual rates of 10.2 and 5.1 percent in the same periods (Islam 1990, cited in Goodman 1991, 2). These rates slowed to an average of 3 percent per year in the period from 1979 to 1988 (Buckley 1990, cited in Goodman 1991, 2).

7. The states were Morelos, Hidalgo, Jalisco, San Luis Potosi, Sonora, and Baja California.

8. Sanderson (1986, 117) reported that, in vegetable processing in Mexico, female and adolescent labor were supplanting former farmers and marginal peasants, "contributing to the exaggerated exploitation of the ejidal household and avoiding the difficulties of sustaining a regular labor force through market incentives."

9. Hiring women is not the only way in which firms seek to combine high-quality labor with lower costs associated with flexibility. The increasing use of contract farming by international agribusiness firms is, in part, an attempt to accomplish this goal by making the contracting family's income dependent on delivering products that meet stringent quality standards and by remunerating per unit of produce, regardless of the amount of labor invested in production (see Watts 1990; Collins 1993). It should be noted that successful contracting also relies on family moralities to discipline its labor force. While contracts are negotiated with the head of the household, they generally draw in the labor of the entire family—labor that as Watts (forthcoming, 67) points out, is in a fundamental sense unfree. This form of family labor, disciplined by the household's stake in output and controlled by the agribusiness firms' specifications of production methods and inputs, mobilizes the social relations of the family to attain production goals.

10. A recent advertisement for Del Monte in *The Packer* touts the company's consistent levels of quality and notes that labor (employees) is the "secret ingredient" in ensuring that quality. The ad features a photograph of two women.

11. Despite the fact that agribusiness firms claim that women require more flexible arrangements, Thomas (1985) found that women in the California lettuce industry had a turnover rate of 20 percent, compared to 75 percent for men (p. 196). Women were more stable with respect to residence and years of tenure with the company and had more years of experience (pp. 184–85).

REFERENCES

Aguiar, Neuma. 1990. *Mujer y Crisis: Respuestas ante la recesión.* Rio de Janeiro: DAWN/Editorial Nueva Sociedad.

Andrade, Manual Correia de. 1984. "Produção de energía e modernização do vale do São Francisco." *Revista de Economía Política* 4, 1: 43–55.

Arizpe, Lourdes, and Josefina Aranda. 1981. "The 'Comparative Advantages' of Women's Disadvantages: Women Workers in the Strawberry Export Agribusiness in Mexico." *Signs* 7, 2: 453–73.

Barrón, Maria Antonieta. 1991. "The Impact of Globalization on the Mexican Labor Market for Vegetable Production." Working paper no. 11. Fresh Fruit and Vegetables Globalization Network, University of California, Santa Cruz, April.

Burbach, Roger, and Patricia Flynn. 1980. *Agribusiness in the Americas,* New York: Monthly Review Press/North American Congress on Latin America.

Collins, Jane L. 1993. "Gender, Contracts and Wage Work: Agricultural Restructuring in Brazil's São Francisco Valley." *Development and Change* 24: 53–82.

Dewalt, Kathleen, and Billie Dewalt. 1992. "The Food Security and Nutrition Impacts of Non-Traditional Agricultural Exports in Latin America." Paper presented to American Anthropological Association. San Francisco, December.

Duque, Ghislaine. 1984. "A experiência de Sobradinho: Problemas fundiarios colocados pelas grandes barragens." *Cadernos de CEAS* (Centro de Estudo e Ação Social) 91: 30–38.

Edwards, Richard. 1979. *Contested Terrain: The Transformation of the Workplace in the Twentieth Century.* New York: Basic Books.

Enloe, Cynthia. 1989. *Bananas, Beaches and Bases.* Berkeley: University of California Press.

Food and Agricultural Organization (FAO). 1985. *Trade Yearbook*, vol. 39.

———. 1989. *Trade Yearbook*, vol. 43.

Gershuny, J. I. 1983. *Social Innovation and the Division of Labour.* Oxford: Oxford University Press.

Goldfrank, Walter. 1991. "Chilean Fruit: The Maturation Process." Working paper no. 16. Fresh Fruit and Vegetables Globalization Network, University of California, Santa Cruz, April.

Gomez, Sergio. 1991. "La uva chilena en el mercado de los Estados Unidos." Working paper no. 18. Fresh Fruit and Vegetables Globalization Network, University of California, Santa Cruz, April.

Goodman, David. 1991. "The Global Fresh Fruit and Vegetable System: Stylized Facts and Some Half-Full Shelves." Working paper no. 1. Fresh Fruit and Vegetables Globalization Network, University of California, Santa Cruz, April.

Gruhn, Isebil V. 1991. "Say It With Flowers." Working paper no. 27. Fresh Fruit and Vegetables Globalization Network, University of California, Santa Cruz, April.

Kessler-Harris, Alice E. 1990. *A Woman's Wage: Historical Meanings and Social Consequences.* Lexington: University Press of Kentucky.

Lago, Maria Soledad. 1987. "Rural Women and the Neo-Liberal Model in Chile." In *Rural Women and State Policy: Feminist Perspectives on Latin American Agricultural Development*, edited by Carmen Diana Deere and Magdalena León, 21–34. Boulder, Colo.: Westview Press.

Lago, Maria Soledad, and Carlota Olavária. 1981. *La participación de la mujer en las economías campesinas: Un estudio de caso en dos comunas frutícolas.* Série resultados de trabajo no. 9. Santiago: Grupo de Investigaciones Agrárias.

———. 1982. "La mujer campesina en la expansión frutícola chilena." In *Debate sobre la mujer en América Latina y el Caribe: Las trabajadoras del agro*, edited by Magdalena León, 179–200. Bogota, Colombia: Asociación Colombiana para el Estudio de la Población.

McAfee, Kathy. 1991. *Storm Signals: Structural Adjustment and Development Alternatives in the Caribbean.* Boston: South End Press.

Mackintosh, Maureen. 1989. *Gender, Class and Rural Transition: Agribusiness and the Food Crisis in Senegal.* London: Zed Books.

Medrano, Diana. 1980. "El caso de las obreras de los cultivos de flores de los municípios de Chia, Cajica y Tabio en la sábana de Bogotá, Colombia." Research paper, Rural Employment Policies Branch, International Labour Organization, Geneva, Switzerland.

Mingione, Enzo. 1991. *Fragmented Societies: A Sociology of Economic Life Beyond the Market Paradigm.* Oxford: Basil Blackwell.

Mohanty, Chandra. 1991. "Cartographies of Struggle." In *Third World Women and the Politics of Feminism*, edited by Chandra Mohanty, Lourdes Torres, and Ann Russo, 1–50. Bloomington: Indiana University Press.

Mones, Belkis, and Lydia Grant. 1987. "Agricultural Development, the Economic Crisis, and Rural Women in the Dominican Republic." In *Rural Women and State Policy: Feminist Perspectives on Latin American Agricultural Development*, edited by Carmen Diana Deere and Magdalena León, 35–50. Boulder, Colo.: Westview Press.

de Oliveira, Orlandina. 1990. "Empleo femenino en México en tiempos de recesión económica: Tendencias recientes." In *Mujer y Crisis: Respuestas ante la recesión*, edited by Neuma Aguiar, 31–39. Rio de Janeiro: DAWN/Editorial Nueva Sociedad.

Padilla, Martha Luz, Clara Murguialday, and Ana Criquillón. 1987. "Impact of the Sandinista Agrarian Reform and Rural Women's Subordination." In *Rural Women and State Policy: Feminist Perspectives on Latin American Agricultural Development*, edited by Carmen Diana Deere and Magdalena León, 124–141. Boulder, Colo.: Westview Press.

Pahl, R. E. 1985. *Division of Labour*. Oxford: Basil Blackwell.

Prates, Suzana. 1990. "Participación laboral femenina en un proceso de crisis." In *Mujer y Crisis: Respuestas ante la recesión*, edited by Neuma Aguiar, 75–92. Rio de Janeiro: DAWN/Editorial Nueva Sociedad.

Pugliese, Enrico. "Agriculture and the New Division of Labor." In *Towards a New Political Economy of Agriculture*, edited by William Friedland, Lawrence Busch, Frederick H. Buttel and Alan Rudy. 137–150. Boulder, Colo.: Westview Press.

Raynolds, Laura T. 1991. "Forces of Instability in Caribbean Participation in International Fresh Produce Systems: Lessons from the Dominican Republic." Working paper no. 21. Fresh Fruit and Vegetables Globalization Network, University of California, Santa Cruz, April.

Rice, Robert A. 1991. "The Globalization of Hybrid Seed Production and the Globalization of Fruits and Vegetables." Working paper no. 2. Fresh Fruit and Vegetables Globalization Network, University of California, Santa Cruz, April.

Roldan, Marta. 1982. "Subordinación genérica y proletarización rural: Un estudio de caso en el Noroeste Mexicano." In *Debate sobre la mujer en América Latina y el Caribe: Las trabajadoras del agro*, edited by Magdalena León, 75–102. Bogotá, Colombia: Asociación Colombiana para el Estudio de la Población.

Ronner, Lucila Díaz, and Maria Elena Muñoz. 1978. "La mujer asalariada en el sector agrícola." *América Indígena* 38: 327–34.

Sáez, Arturo. 1986. *Uvas y manzanas, democracia y autoritarismo: El empresario frutícola chileno (1973–1985)*. Documento de Trabajo no. 30. Santiago, Chile: Grupo de Investigaciones Agrarias.

Sanderson, Steven E. 1986. *The Transformation of Mexican Agriculture: International Structure and the Politics of Rural Change*. Princeton, N.J.: Princeton University Press.

Sarris, Alexander H. 1984. *World Trade in Fruits and Vegetables: Projections for an Enlarged European Community*. Foreign Agricultural Economics Report no. 202. Washington, D.C.: U.S. Department of Agriculture.

Sen, Gita, and Caren Grown. 1987. *Development, Crises and Alternative Visions: Third World Women's Perspectives*. New York: Monthly Review Press.

Serrano, Claudia. 1990. "Mujeres de sectores populares urbanos en Santiago de Chile." In *Mujer y Crisis: Respuestas ante la recesión*, edited by Neuma Aguiar, 93–104. Rio de Janeiro: DAWN/Editorial Nueva Sociedad.

Silva de Rojas, Alicia E., and Consuelo Corredor de Prieto. 1980. "La explotación de la mano de obra femenina en la industria de las flores: Un estudio de caso en

Colombia.'' Research paper, Rural Employment Policies Branch, International Labor Organization, Geneva, Switzerland.

Spindel, Cheywa R. 1987. ''The Social Invisibility of Women's Work in Brazilian Agriculture.'' In *Rural Women and State Policy: Feminist Perspectives on Latin American Agricultural Development*, edited by Carmen Diana Deere and Magdalena León, 51–66. Boulder, Colo.: Westview Press.

———. 1990. ''Mujer y crisis en los años ochenta.'' In *Mujer y Crisis: Respuestas ante la recesión*, edited by Neuma Aguiar. Rio de Janeiro: DAWN/Editorial Nueva Sociedad.

Stolcke, Verena. 1984. ''The Exploitation of Family Morality: Labor Systems and Family Structure on São Paulo Coffee Plantations, 1850–1979.'' In *Kinship Ideology and Practice in Latin America*, edited by Raymond T. Smith. Chapel Hill: University of North Carolina Press.

Thomas, Robert J. 1985. *Citizenship, Gender and Work: Social Organization of Industrial Agriculture*. Berkeley: University of California Press.

Truelove, Cynthia. 1989. ''Flexible Specialization or Specializing in Flexibility: The Industrial Informal Sector in Rural Colombia.'' Paper presented to Program in Comparative International Development, Johns Hopkins University.

Watts, Michael. 1990. ''Peasants under Contract: Agro-food Complexes in the Third World.'' In *The Food Question: Profits versus People?* edited by Henry Bernstein, Ben Crow, Maureen Mackintosh and Charlotte Martino, 149–62. New York: Monthly Review Press.

———. 1994. ''Life Under Contract: Contract Farming, Agrarian Restructuring and Flexible Accumulation.'' In *Living under Contract: Contract Farming and Agrarian Transformation in Sub-Saharan Africa*, edited by Peter D. Little and Michael J. Watts, 21–77. Madison: University of Wisconsin Press.

Wells, Miriam. 1984. ''The Resurgence of Sharecropping: Historical Anomaly or Political Strategy?'' *American Journal of Sociology* 90, 1: 1–29.

Wheelock, Jane. 1990. ''Capital Restructuring and the Domestic Economy: Family Self Respect and the Irrelevance of 'Rational Economic Man.' '' *Capital and Class* 41: 103–41.

12

Depeasantization and Agrarian Decline in the Caribbean

Ramon Grosfoguel

Changes in the world-system in the last four decades have transformed the mode of incorporation of many Caribbean societies in the world-economy from predominantly agrarian export economies to mining, tourism, and manufacturing export economies. This transformation has entailed the abandonment of agriculture rather than its reorientation from exports to production for the domestic market. Changes in the composition of exports and the agrarian decline are the result of general transformations in the peripheral incorporation of the Caribbean to the world-system during the postwar years:

1. Massive flows of direct foreign capital investments in nonagricultural areas from the new core power in the region, namely, the United States (Long 1989; Stone 1983).

2. The formation of nation-states in the region caused partly by the decline of the British empire, which produced the decline of the mercantilist-colonial arrangements that protected the Caribbean agrarian products in the core markets (Stone 1983) and the emphasis of political elites on strategies of industrial development divorced from agriculture (Thomas 1988).

3. The emergence of new technologies in the capitalist world-economy that have substituted artificial sweeteners for the dominant agrarian export of the Caribbean, namely, sugar (Thomas 1985).

These transformations have had important consequences for the agrarian population. If there was one feature that these small Caribbean islands had in common, as of 1945, it was the large proportion of peasants. Today the proportion of the population made up of peasants is rapidly declining; therefore, many Caribbean societies can no longer be characterized as peasant societies (Thomas 1988). Even those Caribbean islands that are still predominantly agrarian have experienced a gradual decline in the economically active population engaged in agriculture. Depeasantization in the Caribbean context is a process of the massive incorporation of underemployed small agrarian producers to wage labor in nonagricultural activities.

This chapter analyzes the processes of depeasantization at two levels: (1) the changing role of the Caribbean region in the international division of labor, referring to the world-system processes mentioned above and (2) peasant households' strategies of coping with declining incomes in agriculture, namely, migrating from rural areas. The post-1945 massive labor export from the Caribbean to the core has intensified the agrarian decline not only through direct or indirect labor mobility away from rural areas but also through private remittances to increasingly nonagricultural rural households. The chapter addresses the role of Caribbean societies within the changing international division of labor, the consequences for agricultural production, and the future developmental options for the region.

Today, following the example of sugar substitution, new technologies have eroded many of the Caribbean's postwar exports including bauxite, nickel, and petroleum, through synthetic substitution. This fact of global capitalist restructuring is seriously affecting the choices facing Caribbean societies in defining their position in the world-economy. The choices are between exclusion (or marginalization) from the global capital flows and incorporation to (or exploitation by) transnational capital flows (Arrighi 1990).

The discussion of these issues is divided into three sections: a historical overview of the features of the Caribbean islands as export-oriented economies; the transformations in Caribbean status in the world-system generating agricultural decline; and the contribution of international migration to this decline.

EXPORT-ORIENTED ECONOMIES

Since the sixteenth century, when Caribbean societies were incorporated as a peripheral region in the capitalist world-economy, the nature of its economies has been export oriented. The Europeans destroyed the indigenous people and populated the islands with African slaves (Mintz 1989). The local economies were reorganized into what came to be known as the plantation system. These plantations were extensions of the European capitalist system to the colonies. The technology and organization of the labor process were transplanted from the European system to the Caribbean colonies (Mintz 1985b). The main difference was the use of slavery in the colonies as the dominant mode to coerce labor in the production of commodities for the European capitalist market (Wallerstein 1980).

The power of the planters was weakened after the slaves were emancipated during the nineteenth century. The Caribbean peasantry came into being when thousands of ex-slaves moved from the plantations and took possession of small plots of land in surrounding areas. Contrary to the European experience, where peasantry was a residual precapitalist system, Caribbean peasants emerged as a result of the capitalist system (Mintz 1989). The refusal of the newly formed Afro-Caribbean peasantry to work in the plantations created a crisis for the planters that lasted throughout the nineteenth century.

By the turn of the century, the sugar plantations had recovered from the postemancipation crisis by reorganizing the labor process either with modern technology or by incorporating East Indian indentured labor. Modern technology was the solution for the Spanish Caribbean where U.S. military interventions in Puerto Rico, Cuba, and the Dominican Republic opened the region up to direct U.S. foreign investments in both banana and sugar plantations (Langsley 1985). East Indian labor was the solution used in the British Caribbean.

With the exception of the Haitian peasants, the Caribbean peasants never developed a subsistence economy or produced enough for the domestic market (Mintz 1985a; Trouillot 1988). Rather, since the late nineteenth century, they were massively tied to agrarian exports. Large multinational sugar refineries or agrarian commercial corporations bought the peasants' sugar or bananas for export to the core markets. As a result, the Caribbean peasants were highly dependent on foreign corporations for their subsistence (Trouillot 1988; Holt 1992). This feature of the Caribbean peasantry is fundamental to understanding its rapid decline as soon as foreign investments in agriculture and agrarian exports declined. After 1945 two transformations in the Caribbean incorporation to the world-system explain this decline.

First, immediately after World War II, the core power in the region shifted from Great Britain to the United States. This shift of power also represented a change in the sectors in which foreign capital investments were concentrated. The U.S. corporations shifted the investments in the Caribbean from agriculture to mining, tourism, and labor-intensive manufacturing, which produced a dramatic transformation in the incorporation of the region to the global division of labor.

Second, as part of the decline of British power, several Caribbean islands began the transition toward independence. A class alliance was formed between the Afro-Caribbean lower strata, descended from the plantation society, and the brown color middle sectors (professionals, small businessman, managers, middle-level public employees) against the colonial system and the white planters (Stone 1983). The alliance was led by the brown color elites under a nationalist discourse. In many instances, this shifted the local balance of power in favor of the brown color elites and significantly weakened the power of the local white planters.

As of the 1960s and 1970s, most of the islands were constituted as nation-states, but their economies were highly dependent on the United States. The elimination of the colonial state corresponded with the gradual elimination of the mercantilist protectionist measures in the core markets for the colonial agrarian products. This change in policy severely affected the residuals of the plantation system, the large estates, whose past reliance on protectionism did not stimulate the planters to upgrade their technology. Thus, the new political elites' economic efforts neglected the uncompetitive large agrarian estates. Constrained by the absence of local capital and the mono-crop economies of the islands, they adapted to the new core power in the region by promoting foreign capital investments in mining, tourism, and industrialization (Thomas 1988). Their policies were consonant with the interests of the new hegemonic power in the region because U.S. transnational corporations exported capital to these new sectors. In sum, the new elites served as intermediaries of foreign capital in the local economy.

These processes transformed the role in the global division of labor of Caribbean societies from specialization in sugar exports to specialization in mining, tourism, and manufacturing. The new modes of insertion accelerated the transformation of most of these societies, gradually decreasing the importance of agriculture. As agrarian foreign capital and the plantation system declined, the local peasants encountered difficulties in exporting their products. No serious effort was made by the postcolonial state to assist them in the marketing of their crops. Instead, the priority of the new political elites was to attract foreign capital to the new, nonagrarian sectors. As a result, underemployed Caribbean peasants migrated to the cities or abroad looking for alternative sources of income in tourist resorts, mines, or manufacturing industries.

CHANGES IN THE CARIBBEAN ECONOMIES

The decline of the agrarian sector in the Caribbean was influenced by the elimination of the protection the colonies had enjoyed within the metropolitan markets after they became nation-states. For centuries, high tariffs in core markets protected the colonies' sugar and bananas from competing peripheral regions. This policy created no incentive for technological upgrading, which, in turn, affected the competitiveness of the region after the colonial mercantilist institutions were eliminated.

The emergence of new artificial sweeteners and corn syrup as a substitute for sugar, the region's main export crop, also affected the decline of agriculture (Thomas 1985). Countries like Jamaica, Guyana, and Barbados, which still had considerable sugar exports, received a serious blow when these substitutes replaced their traditional export. The only country that maintained high levels of sugar production was Cuba because, as part of the Soviet bloc, Cuban sugar enjoyed a highly protected market. However, subsequent to the fall of the Soviet bloc, Cuba has found it difficult to find new outlets for its sugar.

Overall, the relative economic importance of agriculture in the Caribbean islands has decreased significantly during the last three decades. As of 1989, eleven Caribbean islands had a percentage of agriculture to gross domestic product (GDP) of 10 percent or less (Table 12.1). Even in those islands where agriculture is still important, it is declining.[1] For example, the Dominican Republic decreased its share of agriculture to exports in both proportional and

Table 12.1
Percentage Contribution of Agriculture to GDP at Factor Cost (percentages of constant prices)

	1970	1975	1980	1985	1989
Antigua and Barbuda		11.36 (1977)	8.8	4.6	4.5
Bahamas				2	2
Belize	22.6 (1973)	22.7	23.5	22	20
Barbados	15.2	10.5	10.3	9.6	6.6
Cuba	16.7	17	13	8.9	13.7
Dominica		35.3	24.8	26.8	25
Dominican Republic	31	23.5	21.6	22	19.8
Grenada			25.8	22.4	20.4
Guadaloupe	11.5		7.3		
Guyana	24.8	22	23.3	28	28.6
Haiti			33.7	35.25	35
Jamaica			6.3	7.3	7.1
Martinique	8	10.2	6.1		
Monserrat	16	7.8	5.1	3.8	4.1
Puerto Rico	4.3	3.9	3.5	3.5	3
St. Kitts and Nevis		20.5 (1977)	18	12	9.2
St. Lucia		18 (1977)	14.25	18.6	21.3
St. Vincent		19.5 (1977)	13.9	19.5	18.4
Surinam	6.6 (1973)	7.4	8.7	10	10
Trinidad and Tobago	8	6.5	3.6	3.5	3.5

Source: World Table 1992, The World Bank. Baltimore: Johns Hopkins University Press, 1992.

Table 12.2
Percentage Contribution of Agricultural Exports in Total Exports (in dollars)*

	1975	1980	1984	1985	1986	1987	1988	1989
Bahamas	.6 15m	.2 10m	.6 21m	.6	.5	.6	.6	.7 18m
Barbados	57 61m	30 68m	13 46m	12	15.5	26	28	24 44m
Belize		68		54		69	56	
Bermuda	.6 249k	.4 159k	.9 368k	.3	.4	.3	.3	.3 98k
Cuba	93 3.4b	88 4.9b	80 5b	80	84	81	82	80 4.3b
Dominica		38		57		90	79	
Dominican Republic	80 711m	53 515m	65 568m	60	63	54	45	43 392m
Guadaloupe	96 80m	89 95m	84 86m	74	86	78	79	74 111m
Guyana	62 223m	58 221m	26 101m	41	45	49	39	43 225m
Haiti	37 38m	51 112m	32 69m	31	38	25	28	33 53.6 m
Jamaica	26 219m	13 123m	19 139m	24	26	26	25	19 180m
Martinique	72 69m	38 45m	55 85m	65	67	69	65	63 124m
Trinidad and Tobago	6.3 114m	1.9 81m	2.2 48m	2.1	4.2	4.3	5.6	5.8 92m
French Guyana	12 290t	3 750t	6 2m	9	8	8	9	8 4.5m

*The numbers below the percentages are the total amount of exports in dollars.
Note: b = billions of dollars; m = millions of dollars; k = thousands of dollars.
Source: FAO Trade Yearbook 1989, vol. 43, Rome, 1990; FAO Trade Yearbook 1981, vol. 35,
    Rome, 1982; Food and Agriculture Organization of the United Nations, "The State of Food
    and Agriculture," (yearly report from 1980 to 1990).

absolute numbers from 65 percent in 1984 to 43 percent in 1989 (Table 12.2). Nevertheless, the process of agrarian decline is uneven across the region. Whereas Trinidad, Guyana, and Jamaica are highly dependent on the exports of raw materials, Cuba is still highly dependent on sugar exports, Dominica on banana exports, and Haiti on coffee exports. Although Guyana and Jamaica increased their agricultural share of exports in both proportional and absolute numbers during the 1980s, they have not reached their 1950s levels. Moreover, this increase was the result of the collapse of their traditional exports (e.g., bauxite) and the Lomé agreements which stimulated banana exports to the protected European market.

Bananas are the only crop that experienced a production increase during the 1980s. New protectionist measures in the core markets (Harker 1991), such as the Lomé Banana Protocol, made preferential arrangements for Caribbean Community (CARICOM) bananas. The United Kingdom is the major market outlet for this commodity. In the 1980s, three Windward islands, Saint Lucia, Dominica, and Saint Vincent and the Grenadines, were the biggest producers of bananas. These islands doubled their banana production during this period (Cepalc 1991). However, the future of this sector is contingent upon, and vulnerable to, changes in the preferential arrangements with the European Community. An example of their vulnerability is that they cannot compete with the better climates, the more upgraded banana production, and the more competitive prices of Central America (Mintz 1985a; Harker 1991).

The overall decline in agrarian exports has significantly affected the Caribbean peasantry. They were not able to survive as peasants producing export crops with the decline in the agrarian export markets. The region's depeasantization is reflected in the significant decrease of the economically active population in agriculture (Table 12.3). All Caribbean societies, including the most agrarian, have experienced a significant decline in the economically active population in agriculture. Even the absolute numbers of the economically active population in agriculture significantly decreased, with the exception of Haiti, Belize, Cuba, and the Dominican Republic. Cuba and the Dominican Republic require discussion. Although the 1990 figures of the Dominican Republic are as yet unavailable, we can state that the accelerated industrialization of the island has produced a significant decline in the economically active population in agriculture during the 1980s. For example, the agrarian population as a percentage of the total population decreased in the Dominican Republic from 56 percent in 1980 to 39 percent in 1987 (Food and Agriculture Organization 1982, 1991); meanwhile, the manufacturing employment in the export processing zones increased from 18,339 in 1980 to 142,339 in 1992 (Secretaría de Estado de Industria y Comercio 1992). On the other hand, the economically active population in agriculture in Cuba are mainly wage workers given the collectivization of agriculture, which expropriated most of the peasants.

Jamaica is an anomaly in the Caribbean. Although the total active population in agriculture decreased significantly from 300,287 (48.8 percent of the labor force) to 127,708 (18 percent) between 1950 and 1980, by 1990 it had increased

Table 12.3
**Economically Active Population in Agriculture for Caribbean
Countries (percentages)***

	1950	1960	1970	1980	1990
Bahamas		15.5a (8,045)	6.9 (4,791)	5.2 (4,554)	3.9b (4,970)
Barbados		24.3 (22,440)	16.2 (13,621)		4.7c (5,700)
Belize	40.5d (8,238)	39 (10,529)		31.2 (14,745)	
Bermuda			1.4 (393)	1.3 (409)	
Cayman Islands					1.4e (240)
Cuba	41.5 (818,706)		30 (790,356)	22.3f (790,869)	
Dominica		50 (11,693)		31g (7,843)	25.2h (7,700)
Dominican Republic		63 (578,100)	54.7 (755,700)	46i (748,800)	
Grenada		40 (10,893)			14.3j (5,560)
Guadaloupe			32.4k (29,170)	17.2L (18,527)	
Guyana	45.9m (67,454)	34.2 (59,790)		20.3 (48,603)	
French Guyana			18.3n (3,132)	11.4 o (3,706)	
Haiti	83.2 (1,453,891)		61.4p (1,428,755)	57.4q (1,222,859)	57.3 (1,535,444)
Jamaica	48.8r (300,287)	36.1 (236,597)		18s (127,708)	23.1 (244,700)
Martinique		38.8t (35,801)	28.1u (25,150)	7.5v (9,844)	
Netherlands Antilles		1.7 (1,045)	.8w (588)	.3x (320)	
Puerto Rico		23 (136,844)	8.1 (55,094)	3.5 (30,381)	3.6y (39,000)

Decline in the Caribbean • 241

Table 12.3 Continued

St. Kitts-Nevis		46.1 (9,036)			
St. Lucia		48.3 (15,144)		39.5 ----	
St. Vincent and the Granadines		40 (9,954)		25.7 (7,459)	
Surinam		24.8z (19,922)		9.2 (7,459)	
Trinidad and Tobago	27aa (58,767)	19.9 (55,407)	12.5 (36,052)	9.1 (34,244)	10.8 (50,700)
Turks and Caicos Is.				13.9 (405)	
British Virgin Islands			7.3 (294)	5.3 (278)	
US Virgin Islands	18.4 (1,661)	5.4 (610)		1.2 (451)	

*Numbers in parentheses are the total amount of economically active population in agriculture.
Note: a = 1963; b = 1989; c = 1991; d = 1946; e = 1991; f = 1981; g = 1981; h = 1989; i = 1981; j = 1988; k = 1967; L = 1982; m = 1946; n = 1967; O = 1982; p = 1971; q = 1982; r = 1953; s = 1982; t = 1961; u = 1967; v = 1982; w = 1972; x = 1981; y = 1992; z = 1964; aa = 1946.
Source: *1945–1989 Yearbook of Labor Statistics*, Geneva: International Labour Office, 1990; *1992 Yearbook of Labor Statistics*, Geneva: International Labour Office, 1993.

again to 244,700 (23.1 percent). This represented an increase in the active agrarian population during the 1980s of nearly 117,000 (Table 12.3). This is explained by the dramatic decline in the value of bauxite and the slow growth of manufacturing jobs in the export processing zones (just 17,000 by 1988), which stimulated the redevelopment of agricultural exports as a strategy to raise foreign currency to pay the foreign debt (Harker 1991; Hillcoat and Quenan 1991).

The Haitian case is also distinct. Although manufacturing export processing zones boomed during the 1970s and 1980s, it is the Caribbean society with the largest proportion of peasants. Haiti is fundamentally a peasant society. Similarly, Dominica and Saint Lucia have experienced a decline in the number of peasants during the last forty years, but they also still have large numbers of peasants (Mintz 1989a). Moreover, with the Lomé Agreements, banana production has been stimulated in the last decade, revitalizing the peasantry in these small islands (Harker 1991).

Despite the tenuous exceptions represented by Guyana, Jamaica, Haiti, Dominica, and Saint Lucia, the general trend in the Caribbean is toward a depeasantization of the region. The majority of the islands today have replaced agriculture with tourism or manufacturing. Puerto Rico drastically neglected

Table 12.4
Tourism Income/Gross Domestic Product (percentage)

	1987
Antigua and Barbuda	81
St. Kitts-Nevis	55
St. Lucia	46
Bahamas	39
Grenada	38
Barbados	27
St. Vincent	26
Jamaica	21
Dominica	13

Source: CEPAL: Estudio económico de America Latina y el Caribe, 1988.

agriculture as of the early 1950s (Dietz 1986) when the island became a tax-free paradise for U.S. labor-intensive manufacturing. Antigua, Bahamas, and other Lesser Antilles islands are today highly dependent on tourism (Table 12.4); Trinidad depends on petroleum and Barbados on manufacturing.

The decline of agriculture had important consequences for the Caribbean during the 1980s debt crisis years. Debt payments have absorbed a large amount of the local resources. The reduction in foreign exchange revenues from agrarian exports increased the pressure over the local population. In addition, although the Caribbean peasantry has been historically dependent on food exports, they still produced some crops for the domestic market. The decline of the peasantry in the Caribbean reduced the few items that were produced for the local market. Consequently, available foreign currency was used for food imports rather than for technology or machinery. Table 12.5 provides evidence for this assertion in that all countries (except Cuba and Trinidad) increased their absolute amounts of food imports. This process of agrarian decline also has been accelerated by the massive labor migration from the Caribbean to the core, which is the topic of the next section.

MIGRATION AND DEPEASANTIZATION

International migration has been an integral part of the history of Caribbean households since the nineteenth century. In the postemancipation period, intra-

Table 12.5
Food Imports as a Percentage of Total Imports (in dollars)*

	1975	1980	1984	1985	1986	1987	1988	1989
Bahamas	2.1 59m	1.4 80m	3.4 144m	5	5	6	5	5 176m
Barbados	20 43m	13 62m	11 75m	12	12	15	15	14 94.8m
Belize				24 31m	24 29m	22 32m	21 37m	
Bermuda	23 32m	16 48m	15 61m	15	16	14	12	11 60m
Cuba	17 659m	12 923m	11 907m	11	9	9	9	10 868m
Dominica						20 13m	18 16m	16 18m
Dominican Republic	14 99m	10 149m	8 99m	8	10	9	10	10 207m
Grenada			28 16m	25 17m	22 18m	24 21m		
Guadaloupe	22 66m	19 129m	16 98m	17	17	16	15	14 186m
Guyana	11 37m	13 42m	43.7 95m	38	42	45	36	40 90m
Haiti	21 30m	30 84m	18 76m	19	22	25	31	36 112m
Jamaica	16 180m	16 183m	15 168m	12	15	13	14	12 218m
Martinique	18 62m	18 136m	15 102m	16	17	15	15	14 198m
Trinidad and Tobago	8 126m	9 285m	19 357m	20	19	20	16	16 196m
French Guyana	21 15m	14 35m	13 32m	14	14	14	11	9 62m

*The numbers below the percentages are the total amount of imports in dollars.
Note: m = millions of dollars.
Source: FAO Trade Yearbook 1989, vol. 43, Rome, 1990; FAO Trade Yearbook 1981, vol. 35, Rome, 1982; Food and Agriculture Organization of the United Nations, "The State of Food and Agriculture," Rome (yearly report from 1980 to 1990).

244 • Recomposition of Agro-Food Systems

Table 12.6
Caribbean Legal Immigrants to the United States[1]

Years	Amount (3)	Total [including Puerto Ricans] (4)
1891-1900 (2)	33,066	----
1901-1910	107,548	----
1911-1920	123,424	136,643
1921-1930	74,899	105,830
1931-1940	15,502	23,711
1941-1950	49,725	227,882
1951-1960	123,091	573,504
1961-1970	470,213	607,335
1971-1980	741,123	798,343
1981-1990	872,051	1,146,516

[1]Refers to immigrants admitted for permanent residence. The numbers in the middle column are taken from the U.S. Immigration and Naturalization Service, *1990 Statistical Yearbook* (Washington, D.C.: U.S. Department of Justice, 1991), Table 2.
[2]For 1890, 1891, 1895, 1896, and 1897, figures represent immigrant aliens arrived. For 1892, 1893, 1894, 1899, and 1900, figures represent immigrant aliens admitted for permanent residence.
[3]Includes Puerto Ricans only from 1891 to 1904.
[4]For 1911 to 1920 and 1971 to 1989 source used for number of Puerto Ricans was Falcón (1990); for 1921–1970 source used was Johnson (1980); and for 1990 source used was Junta de Planificación de Puerto Rico, *Socio-Economic Statistics* (San Juan: Junta de Planificación, 1990), p. 8.

Caribbean migration became a form of resistance for the emergent peasantry to weaken the power of the local planters and to reconstitute themselves as peasants through the use of remittances (Richarson 1983). After U.S. political and economic interventions in the region early in this century, intra-Caribbean migration gave way to migration from the periphery to the core. This process accelerated during the postwar period (1945–1990), when a massive flow of people migrated from the Caribbean to the United States (Table 12.6). Households sent migrants to core areas as a strategy to cope with poverty and underemployment. The majority of them came from the Greater Antilles, especially from the Spanish Caribbean (Table 12.7).

The Caribbean people have historically moved from peasantization as resistance to resisting peasantization. If in the nineteenth century becoming a peasant was a strategy of resistance against the plantation system, today many house-

Table 12.7
Country Origin and Percentage of Total Caribbean Legal Immigrants per Decade

Years	P.R.	Cuba	D.R.	Jam.	Haiti	Other	Total
1921-30	30,931 29%	15,901 15	----	----	----	58,998 56	100%
1931-40	8,209 35%	9,571 40%	1,150 5%	----	191 1%	4,590 19%	100%
1941-50	178,157 78%	26,313 12%	5,627 2%	----	411 .4%	17,374 7.6%	100%
1951-60	450,413 79%	78,948 14%	9,897 1.7%	8,869 1.5%	4,442 .7%	20,935 3.1%	100%
1961-70	137,122 23%	208,536 34%	93,292 15%	74,906 12%	34,499 6%	58,980 10%	100%
1971-80	57,217 7%	264,863 33%	148,135 19%	137,577 17%	56,335 7%	134,216 17%	100%
1981-90	274,465 24%	144,578 13%	252,035 22%	208,148 18%	138,379 12%	128,911 11%	100%

Sources: U.S. Immigration and Naturalization Service, *1990 Statistical Yearbook* (Washington, D.C.: U.S. Department of Justice, 1991), Table 2; Falcón (1990); Johnson (1980); Junta de Planificación de Puerto Rico, *Socio-Economic Statistics* (San Juan: Junta de Planificación, 1990), p. 8.

holds aspire to break with their peasant past. Migration is a household strategy used by Caribbean peasants to cope with underemployment and low incomes. International migration to the core is an important strategy in this process. Labor mobility from the periphery to the core is the product of household strategies within the constraints of the global division of labor (Portes and Walton 1981). A legitimate outlet to local economic and political oppression today is to send a household member to the metropolis. Migrants' households in the sending countries are in better economic conditions than households without family members in the core (Grasmuck and Pessar 1991; Guarnizo 1992).

Central to understanding the migrant households' social background and the effects of international migration over the peripheral society is the mode of incorporation of the peripheral state into the global interstate system. A major tenet of this chapter is that international migration undercuts the Caribbean agricultural sector in multiple ways depending on the status of the peripheral society: as a colony or a nation-state. On the one hand, Caribbean migrants from nation-states tend to be mainly urban, educated, skilled workers with household incomes that are higher than average (Bray 1984; Grasmuck and Pessar 1991;

Stepick and Portes 1986; DeWind and Kinley 1988; Foner 1983; Portes and Bach 1985; Pedraza-Bailey 1985). On the other hand, Caribbean migrants from societies that still today have or until very recently had a colonial relation with their metropolis tend to be mainly rural and unskilled with low levels of education and from low-income households. For instance, Puerto Rico, Martinique, Guadaloupe, and Surinam have the largest number of migrants to the metropolitan centers as a percentage of the home population (Table 12.8), and their composition is more rural and lower class (Condon and Ogden 1991; Bovenkerk 1987; Freeman 1987; Levine 1987; Centro de Estudios Puertorriqueños 1979). Due to the peculiar colonial status of these islands, the colonial population share the metropolitan citizenship, which gives them privileged access to the metropolis. As a result, they need no visas to enter the core country, making migration more accessible to the poorest sectors of the local population and possible for larger numbers as opposed to the migration from the peripheral nation-states.

Peasants have been the principal source of the migrant flow of colonial societies in the Caribbean. Puerto Rico is the most extreme case, where the agrarian question became obsolete with the exportation of the peasantry to the mainland's urban ghettos. In 1950 agrarian laborers represented nearly 45 percent of the Puerto Rican labor force; by 1980 they represented only about 3.5 percent. Meanwhile, Puerto Ricans living in the mainland represented 75 percent of the island's population as of 1990 (Table 12.8).

The same patterns can be traced in reference to Surinam and Jamaica before independence and the French Antilles to this date, Surinam follows Puerto Rico with the largest numbers of migrants in the metropolis as a percentage of the home population, namely, 49 percent by 1980 (Table 12.8). Most of the migrants from Surinam were rural workers (Bovenkerk 1987). The French Antilles have slightly more than one-quarter of their home population in France[2] (Table 12.8). Most of them were also low-income rural migrants (Cross 1979, 38–39, 70–72).

Jamaica, an interesting case, illustrates the relationship among migration, the colonial insertion, and depeasantization. Before independence, approximately 180,000 Jamaicans migrated to England between 1950 and 1962; most of these were Jamaican peasants (Koslofsky 1981; Foner 1983). In contrast, most of the postindependence Jamaican migrants move to the United States and come from urban, middle-sector backgrounds (Foner 1983). This illustrates the correlation between, on the one hand, rural migration and a colonial status, and, on the other hand, urban, middle-sector migration and a national status.

An important effect of the migration of urban workers from Caribbean peripheral nation-states to core countries is the creation of job opportunities in the urban areas of sending countries which, in turn, fosters the domestic rural-urban migration of many peasants (Sassen 1988). Thus, despite its differences with the migration from a colonial state, international migration from a Caribbean nation-state also produces depeasantization. However, rural-urban peasant migrants in these nation-states were congested in the cities, forming shantytowns around the urban areas, given their restricted access to visas for migrating to the core.

Table 12.8
Caribbean Migrants in the Metropolis

Country	Year	Home Population	Migrants living in the metropolis	Metropolis	Migrants in the metropolis (as a % of the home population)
Puerto Rico (1)	1980	3,196,520	2,014,000	USA	63%
	1990	3,522,037	2,651,815		75%
Surinam (2)	1975	365,000	150,000	Netherlands	41%
	1980	356,000	176,000		49%
Martinique (3)	1982	330,000	95,704	France	29%
Jamaica (4)	1990	2,404,000	435,024 (Ancestry)	USA	18%
			685,024 (includes illegals)		28%
Guadaloupe (5)	1982	332,000	87,024	France	26%
Haiti (6)	1990	6,349,000	289,521 (Ancestry)	USA	4.5%
			689,521 (includes illegals)		10.8%
Dominican Republic (7)	1990	6,948,000	520,151 (Ancestry)	USA	7.4%
			225,000 (includes illegals)		10.7%
Cuba (8)	1983	9,771,000	910,867	USA	9.3%
	1990	10,500,000	1,053,197		10%

[1]U.S. Bureau of the Census, Persons of Hispanic Origin for the United States: 1990, special tabulation prepared by the Ethnic and Hispanic Branch, Washington, D.C.; 1990 Census of Population, General Population Characteristics, Puerto Rico. U.S. Government Printing Office: Washington, D.C.; 1980 Census of Population, General Population Characteristics, Puerto Rico. U.S. Government Printing Office: Washington, D.C.

[2]Rath 1983; Bovenkerk 1987.

[3]Condon and Ogden 1991; Freeman 1987. Migrants living in the metropolis are for foreign born.

[4]U.S. Bureau of the Census, Ancestry of the Population in the United States, special tabulation prepared by the Ethnic and Hispanic Branch, Report CPH-L-89, Washington, D.C.: U.S. Department of Commerce; Maingot 1992.

[5]Condon and Ogden 1991; Freeman 1987. Migrants living in the metropolis are for foreign born.

[6]U.S. Bureau of the Census, Ancestry of the Population in the United States, special tabulation prepared by the Ethnic and Hispanic Branch, Report CPH-L-89, Washington, D.C.: U.S. Department of Commerce; Maingot 1992.

[7]U.S. Bureau of the Census, Ancestry of the Population in the United States, special tabulation prepared by the Ethnic and Hispanic Branch, Report CPH-L-89, Washington, D.C.: U.S. Department of Commerce; Maingot 1992.

[8]U.S. Bureau of the Census, Persons of Hispanic Origin for the United States: 1990, special tabulation prepared by the Ethnic and Hispanic Branch, Washington, D.C.; Maingot 1992.

Domestic or internal migration is an important determinant in the decrease of the economically active population in agriculture and in the decline of agrarian production. Today, the Caribbean is one of the most highly urbanized regions in the world. By 1980 about 52 percent of the total Caribbean population lived in urban areas, which is significantly above the world average of 41.3 percent (Potter 1989). The decline of agriculture, together with the development of tourist resorts, government services, and manufacturing industries in urban areas, fostered a massive rural-urban migration during the 1960s and 1970s, which was reflected in the significant reduction of the peasantry in most of the Caribbean (Potter 1989; Hope 1986; Cross 1979).

Another important analytical distinction is between international migration to core societies and to other Caribbean societies. Each one affects differently the reproduction of the Caribbean peasantry. The main difference is that intra-Caribbean migration tends to support the subsistence of the peasant households, but international migration to core societies is an important source for the eradication of Caribbean peasantries. Intra-Caribbean migration, such as Haitian peasants to the Dominican Republic or Dominican peasants to Guadaloupe, helps these migrants maintain their position as peasants in the sending societies through the transfer of remittances. In contrast, international migration to core countries is more permanent, fostering the migration of entire families. In the cases where some relatives are left behind, the remittances are usually so disproportionate with the average incomes in the sending countries that it is possible to invest in nonagrarian small businesses or in conspicuous consumption. Either way, the money is channeled away from agriculture.

Land prices have increased dramatically due to land speculation of tourist investments and the foreign remittances invested in housing. In a community in the Dominican Republic, land speculation caused by migrant families' remittances produced a 1,000 percent land price inflation over a period of fifteen years (Pessar 1982, 353).

There is little evidence that migration and remittances and savings spur agricultural production and productivity. Land speculation and absentee landholding often contribute instead to a reduction in production. Second, evidence from many Caribbean islands indicates that migrant households cut back on their commercial and subsistence farming activities as they become more reliant on remittances for consumption needs. . . . Third, in the wake of migration, estate owners have frequently found it difficult to find wage laborers. . . . These findings indicate that migration often has negative consequences for agricultural production and productivity. (Pessar 1991, 208)

Thus, such islands as Monserrat, the Dominican Republic, Saint Kitts-Nevis, and Saint Vincent have experienced increased values in land; as a result, landless peasants cannot afford to buy property (Pessar 1991).

CONCLUSIONS

I have argued that the Caribbean economies have been restructured within the ongoing transformations in the world-system. These include especially extension of the states system to the Caribbean and an associated reconfiguring of the international division of labor. The postwar political and economic incorporation of these islands in the world-economy has reduced the significance of agriculture with the expansion of mining, tourism, and light industries. The United States, as the new hegemonic power in the region, changed its foreign capital investments away from agriculture. Migration has been the main peasant household strategy responding to, and accelerating, agrarian decline.

The following questions need to be addressed: Will the trend toward agricultural decline be long-standing or temporary? What are the future possibilities for the region within the world-economy? These questions are important in light of the fact that many of the islands' economies are becoming redundant in the world-economy; bauxite, nickel, sugar, coffee, and other export staples are losing prominence in light of the competition from substitutes created by the new technologies in the core. Even the strategy of attracting foreign industries in manufacturing has not been a successful one for the long-term development of the region. For instance, Puerto Rico successfully attracted foreign industries between 1950 and 1970, but after 1970, the island deindustrialized when industries moved to new peripheral areas, such as Haiti and the Dominican Republic, which offered cheaper labor and lower costs of production (Bonilla and Campos 1986). Unemployment in Puerto Rico rocketed. Is this the future awaiting the Dominican Republic? The probability is that, as Dominican workers' salaries increase, other regions with cheaper labor could become more attractive to foreign capital investors.

Although the "industrialization by invitation" strategy has been successful during the 1980s in the Dominican Republic, some countries, such as Jamaica and Barbados, have been unsuccessful in their attempts to attract foreign capital in manufacturing industries. Competition from semiperipheral regions, such as Mexico's northern border, have affected the Caribbean. Overall, the expectations of industrialization created by the local elites in the postwar era were never fulfilled in the Caribbean.

One possible scenario is a comeback of agriculture. Several islands during the 1980s experienced an increase in agricultural production, after several decades of decline, as a result of preferential arrangements with the European Community. However, unless the Caribbean banana producers take advantage of this opportunity to upgrade their technology, the growth of the banana industry will last as long as the protectionist measures exist. A higher technological level could make Caribbean bananas more competitive in core markets vis-à-vis the Central American bananas once the protectionist measures are eliminated. This scenario is limited by the chronic debts of many Caribbean societies. It is extremely difficult for highly indebted countries to acquire technology in the international markets. Alternatively, these islands could attempt to attract transnational agribusiness corporations.

Despite the uncertainties of the future, Caribbean societies are struggling today to redefine their insertion in the capitalist world-economy. The choice today is similar to that identified by Arrighi (1990) in relation to other peripheral societies, that is, exclusion versus exploitation. In recent years, the general trend has been toward the exclusion of the Caribbean's traditional export items from global markets because of their replacement by high-tech substitute products. Ironically, today, many Caribbean islands would prefer to be exploited by, rather than be excluded from, transnational capital flows. The Caribbean elites are developing strategies against these exclusionary tendencies. As Arrighi (1990) rightly predicts, this is leading them to "seek reentry into the world division of labor on conditions favorable to core states" (p. 17), fostering the conditions for a more exploitative reinsertion in the world-economy. The contradictory relationship between exclusion and exploitation usually plays in favor of the exploitation of the periphery by the core. The Caribbean, as a peripheral region, is no exception to this pattern.

NOTES

1. The perception of the importance of agriculture varies according to whether it is seen as a percentage of the total exports or as a percentage of the GDP; the former ascribes greater importance to agriculture than the latter because tourism and manufacturing in the export processing zones are not included in the total amount of exports (Table 12.2). For example, in such countries as Martinique, Guadaloupe, Barbados, Saint Kitts-Nevis, and Jamaica, agriculture as a percentage of total exports is high, but as a percentage of the GDP it is low. Thus, it is misleading to use agriculture as a percentage of total exports as an indicator of whether these economies are agrarian; rather, it is preferable to look at agriculture as a percentage of the GDP to account for the proportional importance of this sector in relation to tourism, export processing zones, and so on. This is significant in light of the fact that many Caribbean islands, especially the smaller ones, have predominantly tourist economies (Table 12.3).

2. This figure counts only those born in the French Antilles. The estimate for the number of people with French Caribbean ancestry in France is around 200,000 per island. This would raise the numbers of migrants as a percentage of the home population to around 60 percent.

REFERENCES

Arrighi, Giovanni. 1990. "The Developmentalist Illusion: A Reconceptualization of the Semi-Periphery." In *Semiperipheral States in the World-Economy*, edited by William G. Martin, 11–42. Westport, Conn.: Greenwood Press.
Bonilla, Frank, and Ricardo Campos. 1986. *Industry & Idleness*. Hunter College of the City University of New York: Centro de Estudios Puertorriqueños.

Bovenkerk, Frank. 1987. "Caribbean Migration to the Netherlands: From Elite to Working Class." In *The Caribbean Exodus*, edited by Barry B. Levine. New York: Praeger.

Bray, David. 1984. "Economic Development: The Middle Class and International Migration in the Dominican Republic." *International Migration Review* 18, 2: 217–36.

Centro de Estudios Puertorriqueños. 1979. *Labor Migration under Capitalism: The Puerto Rican Experience*. New York: Monthly Review Press.

Cepalc. 1991. *Anuario Estadístico*. Santiago, Chile: Naciones Unidas.

Condon, Stephanie A., and Phillip E. Ogden. 1991. "Emigration from the French Caribbean: the Origins of an Organized Migration." *International Journal of Urban and Regional Research* 15, 4 (September): 505–23.

Cross, Malcolm. 1979. *Urbanization and Urban Growth in the Caribbean*. Cambridge, England: Cambridge University Press.

DeWind, Josh, and David H. Kinley III. 1988. *Aiding Migration: The Impact of International Development Assistance on Haiti*. Boulder, Colo.: Westview Press.

Dietz, James L. 1986. *Economic History of Puerto Rico*. Princeton, N.J.: Princeton University Press.

Falcón, Luis Nieves. 1990. "Migration and Development: The Case of Puerto Rico." Economic Development working paper no. 18. Wilson Center, Washington, D.C.

Foner, Nancy. 1983. "Jamaican Migrants: A Comparative Analysis of the New York and London Experience." Occasional paper no. 36. New York University, Center for Latin American and Caribbean Studies.

Food and Agriculture Organization (FAO). 1982. *The State of Food and Agriculture 1981*. Rome: Food and Agriculture Organization of the United Nations.

———. *The State of Food and Agriculture 1990*. Rome: Food and Agriculture Organization of the United Nations.

Freeman, Gary P. 1987. "Caribbean Migration to Britain and France: From Assimilation to Selection." In *The Caribbean Exodus*, edited by Barry B. Levine, 185–203. New York: Praeger.

Girvan, Norman. 1989. "Technological Change and the Caribbean: Formulating Strategic Responses." *Social and Economic Studies* 38, 2: 111–35.

Grasmuck, Sherri, and Patricia Pessar. 1991. *Between Two Islands: Dominican International Migration*. Berkeley: University of California Press.

Guarnizo, Luis E. 1992. *One Country in Two: Dominican-owned Firms in New York and the Dominican Republic*. Ph.D. diss. Johns Hopkins University.

Harker, Trevor. 1991. "The Impact of External Sector Developments on Caribbean Economic Performance 1983–88." *Caribbean Studies* 24, 1–2: 1–44.

Hillcoat, Guillermo, and Carlos Quenan. 1991. "International Restructuring and Respecialization of Production in the Caribbean." *Caribbean Studies* 24, 1–2: 191–222.

Holt, Thomas C. 1992. *The Problem of Freedom: Race, Labor, and Politics in Jamaica and Britain, 1832–1938*. Baltimore: Johns Hopkins University Press.

Hope, Kempe Ronald. 1986. *Urbanization in the Commonwealth Caribbean*. Boulder, Colo.: Westview Press.

Johnson, Roberta Ann. 1980. *Puerto Rico: Commonwealth or Colony?* New York: Praeger.

Koslofsky, J. 1981. "Going Foreign: Causes of Jamaican Migration." *NACLA* 15, 1: 2–31.

Langsley, Lester D. 1985. *The United States and the Caribbean in the Twentieth Century*. Athens: University of Georgia Press.

Levine, Barry B. 1987. "The Puerto Rican Exodus: Development of the Puerto Rican Circuit." In *The Caribbean Exodus*, edited by Barry B. Levine, 93–105. New York: Praeger.

Long, Frank. 1989. "Manufacturing Exports in the Caribbean and the New International Division of Labor." *Social and Economic Studies* 38, 1: 115–31.

Maingot, Anthony P. 1992. "Immigration from the Caribbean Basin." In *Miami Now!*, edited by Guillermo J. Grenier and Alex Stepick III, 18–40. Gainsville: University Press of Florida.

Mintz, Sidney W. 1985a. "From Plantations to Peasantry in the Caribbean." In *Caribbean Contours*, edited by Sidney W. Mintz and Sally Price, 127–154. Baltimore: Johns Hopkins University Press.

———. 1985b. *Sweetness and Power*. New York: Penguin Books.

———. 1989. *Caribbean Transformations*. New York: Columbia University Press, Morningside Edition.

Pedraza-Bailey, Silvia. 1985. *Political and Economic Immigrants in America: Cubans and Mexicans*. Austin: University of Texas Press.

Pessar, Patricia. 1982. "The Role of Households in International Migration and the Case of U.S.–Bound Migration from the Dominican Republic." *International Migration Review* 16, 2: 342–62.

———. 1991. "Caribbean Emigration and Development." In *The Unsettled Relationship*, edited by Demetrios G. Papademetriou and Philip L. Martin, 201–210. New York: Greenwood Press.

Portes, Alejandro, and Robert L. Bach. 1985. *Latin Journey: Cuban and Mexican Immigrants in the United States*. Berkeley: University of California Press.

Portes, Alejandro, and John Walton. 1981. *Labor, Class and the International System*. New York: Academic Press.

Potter, Robert B. 1989. "Urbanization, Planning and Development in the Caribbean: An Introduction." In *Urbanization, Planning and Development in the Caribbean*, edited by Robert B. Potter, 1–18. London: Mansell.

Rath, John. 1983. "Political Participation of Ethnic Minorities in the Netherlands." *International Migration Review* 17, 3: 445–69.

Richarson, Bonham C. 1983. *Caribbean Migrants: Environment and Human Survival on St. Kitts and Nevis*. Knoxville: University of Tennessee Press.

Sassen, Saskia. 1988. *The Mobility of Labor and Capital: A Study in International Investment and Labor Flow*. London: Cambridge University Press.

Secretaría de Estado de Industria y Comercio. 1992. "Empresas de zonas francas y empleos generados." Santo Domingo, Dominican Republic: Consejo Nacional de Zonas Francas de Exportación.

Stepick, Alex, and Alejandro Portes. 1986. "Flight into Despair: A Profile of Recent Haitian Refugees in South Florida." *International Migration Review* 20 (Summer): 329–50.

Stone, Carl. 1983. "Patterns of Insertion into the World-Economy: Historical Profile and Contemporary Options." *Social and Economic Studies* 32, 3: 1–34.

Thomas, Clive Y. 1985. *Sugar: Threat or Challenge: An Assessment of the Impact of*

Technological Developments in the High Fructose Corn Syrup and Sucrochemical Industries. Ottawa, Canada: Ottawa International Development Research Centre.

————. 1988. *The Poor and the Powerless.* New York: Monthly Review Press.

Trouillot, Michel-Rolph. 1988. *Peasants and Capital: Dominica in the World Economy.* Baltimore: Johns Hopkins University Press.

Wallerstein, Immanuel. 1980. *The Modern World System.* Vol. 2, *Mercantilism and the Consolidation of the European World-Economy, 1600–1750.* New York: Academic Press.

13

Free Trade Agreements: Their Impact on Agriculture and the Environment

Robert Schaeffer

The free trade agreements (FTAs) now being negotiated, signed, and deployed do not promote free trade.[1] Instead they protect core countries and extend the power of transnational corporations at the expense of consumers, taxpayers, domestic producers, and the environment in core and peripheral states. FTAs may expand the volume of trade, but there is little evidence that expanded trade provides wider economic opportunities or improves the fortunes of disadvantaged economic actors. FTAs deserve critical attention because they contribute to the emergence of a new economic order, what I would call "indifferent imperialism," which may take a global or regional form depending on which free trade agreements are signed and ratified by member states.

Since 1947 member states have used the General Agreement on Tariffs and Trade (GATT) and its periodic renegotiation process (called "rounds") to reduce tariffs and promote the uniform economic treatment of its members, numbering 112 in 1933. In the mid-1980s, however, the United States departed from the GATT negotiating framework and adopted a complex negotiating strategy in multiple forums. It abandoned its long-standing focus on lower tariffs and uniform treatment and introduced a new, wider economic agenda designed to defend its fading trading position in the world.

In strategic terms, U.S. officials initiated a two-track process to negotiate more favorable trade relations. The first track consisted of essentially bilateral trade negotiations between the United States and other countries. U.S. trade officials used Section 301 of the Trade Acts of 1984 and 1988 as a retaliatory "crowbar" to force open trade negotiations with countries that denied U.S. corporations "reasonable" access to domestic markets, dumped goods in the United States at below-market prices, or failed to protect the patents and copyrights (called "intellectual property rights" by U.S. trade officials) of U.S. firms (Davidow

1991, 47; Raghavan 1990, 73; Bradsher 1992, C1; Bhagwati 1989, C2). The threat of retaliation was used to force peripheral states to comply with U.S. trade demands in bilateral negotiations and in multilateral forums like GATT.[2]

In addition, U.S. officials negotiated and concluded bilateral free trade agreements, first with Canada in 1988 and then with Mexico in 1992 (which has been signed but not ratified). Administration officials used these bilateral agreements both as a model for the new round of GATT negotiations that began in 1986— North America Free Trade Agreement (NAFTA) provisions were more radical and comprehensive than GATT proposals—and as a fallback position should GATT negotiations fail to achieve U.S. policy objectives.

The second track of the U.S. negotiating strategy was to use the "Uruguay Round" GATT negotiations as an opportunity to introduce a new set of issues that went far beyond GATT's traditional focus on lower tariffs and uniform treatment.[3] Successive GATT rounds had reduced average tariffs on manufactured goods from about 40 percent in 1950 to about 5 percent in 1990, which meant that the original aims of GATT had been substantially achieved (Orr 1992, 118). Indeed, the economic gains that could be realized by the complete elimination of tariffs were quite modest, worth perhaps $5 billion in exports to the United States, even less to Japan (Orr 1992, 118–19; Jerome 1992, 172). Because tariffs were already low, U.S. officials introduced a new agenda, proposing that agricultural tariffs and subsidies be discussed for the first time. They proposed eliminating government subsidies and regulations that acted as "nontariff barriers to trade," protecting corporate holders of patents and copyrights; banning restrictions on the import or export of food, fiber, energy, and natural resources; and making countries treat multinational corporations and some professional workers in the same way that they treated domestic firms and citizens.

In addition, U.S. officials used debt crisis negotiations, which were then being conducted by multilateral institutions, such as the World Bank, on behalf of private banks and donor states with heavily indebted countries, to press for many of the same goals: the reduction of tariffs and restrictions on foreign investment, the privatization of state assets, the devaluation of currencies so that privatized assets could be cheaply purchased by foreign investors, and the elimination of government subsidies to domestic producers and consumers (Schaeffer 1993, 176).

In Mexico, for instance, debt crisis negotiations in 1982 led to the wholesale reduction of average tariffs from 40 to about 10 percent. NAFTA would reduce these already low tariffs to 0 percent over the next ten years.[4] With the exception of the oil industry, many state assets were privatized, and the devaluation of the peso—from 25 pesos to the dollar in 1982 to 3,100 pesos to the dollar in 1992— has encouraged the acquisition of Mexican assets by U.S. firms and widespread investment throughout Mexico.[5] In this context, NAFTA represents the endgame consolidation of changes made in response to debt crisis negotiations.[6]

Taken together, the U.S. strategy can be described as a two-ring circus, one in which U.S. negotiators have used different forums to press for similar measures under the big-top rubric of free trade. Although they risk collision, the different FTAs are joined by a common set of principles and objectives designed to shore up the U.S. trading position, which had weakened considerably in the 1980s.

THE AGRICULTURAL PRELUDE

In 1978 the Iranian-led oil embargo forced oil prices up. Since oil is the most important U.S. import,[7] rising oil prices increased the cost of imports, which worsened the U.S. balance of trade and also led to inflation. To combat inflation, and to raise money for military expansion without raising taxes to pay for it, U.S. officials in 1979–1980 raised interest rates. This led to large and growing federal budget deficits—$2.5 trillion between 1980 and 1992 (Skidmore 1992, 16)—and triggered a crisis both for indebted peripheral countries and indebted American farmers, who had borrowed heavily to purchase land and expand production in the 1970s.[8] Indebted states could not afford to purchase imported goods, so demand for U.S. agricultural goods contracted. U.S. farmers, meanwhile, intensified production to meet debt obligations, or went bankrupt in the attempt, further increasing agricultural supplies.

At the same time, farmers in the European Community steadily increased their production of agricultural goods and sold mounting surpluses on overseas markets. Burgeoning core-country supplies and falling peripheral-country demand for imports, which was a product both of indebtedness and the increasing self-sufficiency of such countries as India,[9] led to falling prices. For the United States, agricultural exports declined from $41 billion in 1980 to $29 billion in 1985 (Hufbauer 1989, 162), and its share of the overseas wheat market shrank from 55 to 36 percent between 1980 and 1986 (Watkins 1991, 40).[10]

The Reagan administration viewed agricultural exports as extremely important because they accounted for one-fifth of all exports and recorded substantial trade surpluses (agricultural, chemical, and high-tech industries were the only sectors that did). In addition, agriculture was more dependent on exports than any other industry, exporting one-third of its cereal production, which accounted for half of total world trade.

To boost exports and reduce growing trade deficits, the Reagan administration used the 1985 Farm Bill to cut loan rates to farmers so that U.S. exporters like Cargill could purchase and sell lower priced goods on overseas markets and regain market shares.[11] But because lower loan rates meant higher deficiency payments to farmers (the government pays the difference between the loan rate and the target price, which is supposed to cover the costs of production—it does for large farmers but not for smaller family farmers), the cost of farm subsidy programs increased to $30 billion in 1986, ten times the 1980 level (Watkins 1991, 40), which added to the growing budget deficit.[12]

In the short run, the Reagan administration was willing to use subsidies to lower prices and wrestle over market shares with the European Community

(EC), which increased its own subsidies to stay competitive. U.S. agriculture secretary John Block described this strategy as "squeezing the CAP [Community Agricultural Policy] until the pips squeak" (Watkins 1990, 2). But, in the long run, U.S. officials decided that if U.S. and EC agricultural subsidies could be eliminated through trade agreements, they could reduce the federal deficit, undercut EC competitors, recapture overseas markets, increase U.S. exports, and improve the U.S. balance of trade. It was in this context that the Reagan and Bush administrations embarked on multitrack, free trade negotiations.

The United States was joined in its efforts to expand the GATT agenda by the European Community, Japan, and a group of agricultural-export states (the Cairns group).[13] They wanted to push for measures that would strengthen transnational corporations, particularly those in the "service" industries, and make government regulations uniform so that production could be standardized on a global basis. Although they agreed on the objectives of this new agenda, Japan and the EC countries disagreed with the U.S.–Cairns group proposals to eliminate agricultural subsidies. They did not necessarily oppose reducing subsidies, which would reduce government payments to farmers and lower the cost of food for workers employed by corporations. Indeed, officials in some states covertly welcomed radical U.S. proposals so that they could give in to outside pressure rather than introduce unpopular measures themselves.[14] The problem was that the centrist-conservative governments in core states that supported the new free trade agenda themselves relied for political support on conservative rural farmers who had long been beneficiaries of government subsidy programs. And unlike in the United States, where small family farmers had been ruined by the farm-debt crisis of the early 1980s, small farmers in Japan and the EC remained economically viable and politically strong. The United States could make radical agricultural proposals because it was both economically advantageous and politically possible.[15]

THE IMPACT OF FTA PROVISIONS ON AGRICULTURE AND THE ENVIRONMENT

Although the provisions of various FTAs vary, it is possible to make a preliminary assessment of their impact by examining existing case histories and analyzing the economic relations affected by different provisions. In general, FTAs will probably have four important consequences: They will promote the monopoly power of transnational corporations; reduce sensible regulations that protect consumers and the environment; cut prices of agricultural goods and natural resources, hurting small producers in core and peripheral states; and guarantee the supply of food and natural resources for producers and consumers in core countries.

Promote Monopoly Power

The FTAs promote the monopoly power of private transnational corporations (TNCs) by facilitating cross-border investment, cutting corporate taxes, promoting U.S. control of agricultural export sectors in other countries, reducing competition from state-subsidized firms, and extending corporate property rights.

By granting TNCs the same rights and privileges of domestic firms, and by eliminating restrictions on profit-repatriation, domestic ownership, and technology transfer requirements, FTAs make it easier for TNCs to make cross-border investments. Although there is already substantial cross-border investment in Mexico, U.S. corporations are making cross-border investments and strategic alliances in anticipation of NAFTA ratification.

Because so much of the world-economy is controlled by TNCs, for whom "export" and "import" are really intrafirm transfers across borders, tariff reduction is really a gigantic reduction of corporate taxes.[16] In North America, where about 40 percent of the Mexican and 30 percent of the Canadian export economy is controlled by U.S. TNCs (1993, 51),[17] tariff reduction provides important tax savings for U.S. TNCs operating in Canada and Mexico. Based on 1990 trade figures, I estimate that the U.S. government will lose about $4 billion, the Canadian government $3 billion, and the Mexican government $3 billion in tax revenues, and multinational firms will capture most of the benefits (Bradsher 1991, 10). And if one includes the costs of retraining U.S. workers who lose their jobs and building a new infrastructure along the U.S.–Mexican border, NAFTA could cost the U.S. government as much as $40 billion (Bradsher 1993a, C2).

Free trade proponents argue that lower tariffs will mean lower consumer prices. But this is a dubious proposition since the corporations that dominate export trade can use their tax savings to pay higher salaries or stock dividends or make new investments rather than offer lower prices to consumers.[18] Even if tariff reduction provided lower prices for consumers, taxpayers would still have to make up lost revenues, at a time when all three governments are strapped for revenue and when governments must still provide border and customs controls to prevent the cross-border migration of people and drugs.

U.S. officials argue that FTAs will promote competition, and they point to NAFTA provisions that eliminate "monopolies." But antimonopoly provisions are aimed at state "subsidized" public enterprises. Private TNCs are not subject to antitrust provisions. In this context, demonopolization does not increase economic competition but enables TNCs to "compete" in areas previously denied to them without the obligation to provide services and benefits once provided to consumers by state-owned or subsidized monopolies.

For the first time, FTA provisions guarantee and extend to patent holders the kind of monopoly protection they now enjoy in core countries. Because increased protection for "intellectual property rights" could increase U.S. patent-holder revenues by between $43 and $102 billion a year, according to some estimates (Rongead 1990, 14; Chomsky 1993, 412), they are a central objective of TNCs and core-country negotiators.

FTAs extend the duration of patents from seventeen to twenty years and of copyrights from seventeen to fifty years. According to Deardorff,

As its name suggests, intellectual property "protection" is a surprising issue for the GATT . . . on issues of commercial policy, the GATT's mission has always been to prevent, or at least circumscribe, countries' efforts to "protect" their domestic industries. Now in the TRIPs area, the GATT is being called upon to *extend protection*, not restrict it. (1990, 498, emphasis added)

The United States, Japan, and the EC are keen on this issue because multinational firms want to clamp down on the piracy of trademarks, patents, and copyrights by other countries. And because core industries hold the overwhelming majority of the world's patents—a product of their advanced educational and industrial research and development infrastructures—patent protection will benefit core countries and strengthen monopoly power (Raghavan 1990, 123).

There are a number of problems with extending patent protection. Consumers in the periphery will be forced to pay higher prices for generic drugs, which are now cheaply produced by domestic pharmaceutical companies. For agriculture, the protection of hybrid seeds and pesticides, the products of biotechnology engineering and food technologies (from food irradiation to invisible bruise technology), will increase core agricultural industry power and increase farmer costs and dependency on the core. Under new rules, farmers would have to pay royalties on patented seeds and pay each time the seeds were used, even if they were saved from the previous harvest.[19]

The irony is that, in agriculture, many of the seeds patented by core-country scientists are culled from plants collected in peripheral countries. "A wild tomato variety taken from Peru in 1962 has contributed $8 million a year to the American tomato processing industry by increasing the content of soluble solids," notes Shiva (1991, 25). According to some sources, peripheral-country germplasm has contributed $66 billion to the U.S. economy (Shiva 1991, 25). But with enforceable patents in hand, core-country TNCs would be able to profit from the sale of these materials to countries where they originated (Watkins 1992, 37–38).[20]

Reduce Consumer and Environmental Protection

A second important feature of FTAs is that they seek to eliminate or reduce "non-tariff trade barriers," arguing that regulations protecting consumers, workers, and the environment obstruct trade. To achieve this, they would deprive public officials of the right to regulate trade and assign this authority to secret arbitration panels or bureaucratic forums, which set lower standards (Schaeffer, forthcoming). The "flexible" global production system, now being created by TNCs, requires a uniform regulatory environment. The elimination or harmonization of regulations and standards facilitates this development.

There is already considerable evidence that the elimination of nontariff trade barriers undermines many sensible measures that protect consumers and the environment from pernicious cross-border trade.[21] Provisions in GATT and the U.S.–Canada FTA have permitted traders to seek the elimination of regulations designed to prevent the sale of diseased puppies by U.S. puppy mills to Canadian pet stores (Farnsworth 1992, C1) and allowed the Mexican tuna-fishing fleet to conduct a successful GATT suit against the U.S. Marine Mammal Protection Act, which barred Mexican tuna shipments because Mexican fishermen slaughtered dolphins in the process (Christensen and Geffin 1991–1992; Christensen 1991).

As a consequence of these rulings and new FTA provisions, it will become increasingly difficult to restrict trade in products on the basis of harvest or manufacturing process. States will be unable to "discriminate" against products manufactured in ways that harm the environment or violate worker rights.[22]

Because public officials in local, state, and national legislatures now have the authority to regulate worker safety and consumer and environmental protection, they have created different levels of "protection."[23] One objective of FTA provisions is to eliminate the government regulations that act as nontariff trade barriers. But where they cannot eliminate regulations, they seek to "harmonize" them or make them uniform. To do this, FTAs seek to deprive public officials of the right to regulate trade and assign this authority to FTA arbitration panels to bureaucratic forums selected because they are insulated from political pressure and public scrutiny.

New language in the GATT requires contracting parties to "take such reasonable measures as may be available to it to ensure observance of the provisions of this agreement by the regional and local governments and authorities within its territory" (Grimmet 1991, 38). And NAFTA would require members to "take all necessary steps, where changes to domestic laws will be required to implement their provisions . . . to ensure conformity of their law with these agreements" (Wallach 1991, 2; Compa 1992, 52).[24] This would greatly undermine the authority of democratic institutions.[25]

The authority to regulate would be assigned either to arbitration panels, of the kind that ruled against puppy- and dolphin-protection legislation, or to global bureaucratic agencies, such as Codex Alimentarius. Codex, an obscure agency of the Food and Agriculture Organization of the United Nations, based in Rome, sets standards and regulations for trade in animal and food products. Whereas adoption of its recommendations is now voluntary, FTAs would assign Codex responsibility for setting uniform global standards. This is a problem because historically Codex sets standards that are lower than those that apply in the United States and the EC. For example, Codex standards for DDT pesticide residues in imported bananas are fifty times higher than the U.S. Environmental

Protection Agency (EPA) permits (Ritchie 1990a, 216). According to Bredahl and Forsythe (1989, 196–97), the EPA accepts only about 20 percent of Codex standards for residual pesticide limits.

The elimination of many nontariff trade barriers and the creation of a uniform regulatory environment make it possible for TNCs to construct more flexible commodity production systems.

Cut Prices

A third objective of FTAs is to cut agricultural prices by ending farm subsidies and supply-management programs. If subsidies were eliminated or reduced, core states could greatly reduce their spending. The United States spends between $20 and $40 billion on subsidies ranging from deficiency payments to soil conservation programs and land set-asides to research at land-grant universities (Soden 1988); the EC spends about $40 billion, nearly two-thirds of its total budget (Tarditi et al. 1989, 2; Johnstone 1992, 24).

The end of government subsidies and supply-management programs would force farmers to intensify production on existing land and, in some cases, expand production on now-idle land (Ritchie 1989, 9–11; 1987a, 9). Increasing production will create larger supplies, which will lower the prices merchant-exporters and industrial food processors pay for produce, and it will increase the sales of agrochemical and machinery producers, which sell inputs that can increase productivity.[26] Increased intensity will also result in greater environmental degradation (Greenpeace 1992). For farmers in core and peripheral states, who lack the capital to intensify production or develop significant economies of scale, increasing supplies and lower prices will mean bankruptcy and displacement. And the end of supply-management programs will force others out of business.

Although FTA proponents argued that reduced subsidies will benefit taxpayers and lower prices will benefit consumers, the control of food supplies by agro-industrial suppliers and merchant-exporters, food processors, and retailers means that the benefits of lower prices are captured primarily by intermediaries.[27] "Between 1981 and 1987, consumer food costs rose 36 percent," at a time when prices paid farmers "declined precipitously," a development DeLind (1992, 2–3) attributes to control over the food supply by merchants and manufacturers.[28]

The reduction of subsidies could also have important environmental consequences. Depending on whether they are defined in FTAs as "trade distorting" or not, "subsidies" that promote soil or water conservation, food inspection, education and cooperative extension, farm credit and loan programs, foreign food aid, and research and development at land-grant universities could all be subjected to attack (Ritchie 1988, 3; Greenpeace 1992).[29] If the grants were eliminated, one consequence could be a shift from public to private agricultural research, both because funding for public-sponsored applied research was reduced (basic research might remain) and because private companies, armed with expanded and enforceable patent rights, would want to capture the benefits of research for themselves, rather than share it widely (Schaeffer, forthcoming).

Of course, because EC farmers are heavily dependent on farm subsidies, even a modest reduction will take a heavy toll on the continent's small farmers and its rural communities. And by increasing cheap core-country grain supplies, the end of subsidies and supply management will contribute to a reduction in peripheral country grain self-sufficiency, which has been on ongoing problem since the advent of the "green revolution."

Guaranteed Supplies

A final element of FTAs, which has important consequences for agriculture and the environment, are provisions guaranteeing supplies of food, agricultural products, and natural resources for TNCs and core countries.

FTA provisions would prohibit member states from restricting food or resource imports, even if these imports were being dumped at low prices to capture market shares.[30] This is a particularly serious problem for peripheral countries, which have already seen domestic grain production fall because imported core-country grains are cheaper (Watkins 1989, 13).

The abolition of import restrictions affects not only producers in peripheral countries, but farmers in core states as well. U.S. beef, sugar, peanut, and tobacco farmers all benefit from import restrictions. If U.S. beef import restrictions were eliminated, production would shift to Central and South America, where the clearing of rain forests to grow pasture for cattle has serious environmental consequences.

The abolition of import restrictions is an extremely important issue for Japan, which has long provided massive subsidies to domestic rice farmers and has banned rice imports. The government has wanted to reduce farm subsidies and import controls to reduce its expenses, provide cheaper food supplies to urban workers, thereby reducing the wage bill for Japanese industry, and augment rice supplies, which are periodically in short supply (Pollack 1992, 1). The government has not done so because it is reluctant to surrender food security and because the ruling party depends on rural farm constituencies for important electoral support, which was the product of a constitution designed with U.S. assistance to ensure conservative party success by giving disproportionate electoral weight to rural voters.

In this context, U.S. officials have offered a grand compromise designed to win Japanese support for GATT. U.S. negotiators have proposed that restrictions on food and natural resource *exports* also be prohibited by FTAs.[31] Countries that could export food to Japan would be required to do so, even if they faced critical domestic shortages of their own. U.S. officials have demanded that members "agree to modify GATT specifically in order to remove the ability of countries to restrict exports in times of critical shortages" (Ritchie 1988, 3). In this context, food security for Japan, and for other core countries that have the capacity to pay for food in difficult times, is guaranteed by undermining the food security of other countries. This strategy allows the United States to neutralize Japan and go after EC subsidies separately.

By preventing countries from restricting food, energy, and natural resource exports, FTAs undermine attempts to establish collective production-marketing arrangements that try to control supplies and raise prices, such as OPEC or the African coffee exporters group. Some FTAs go so far as to prohibit members from reducing exports below the average of the previous three years, a provision that essentially establishes export *quotas*. The NAFTA provisions are designed to guarantee supplies of Canadian natural gas, water, and electricity and Mexican oil to the United States (Shrybman 1990, 20; Tester 1988, 204–13).

The elimination of export restrictions also limits the ability of states to manage sustainably their natural resources. Many members of the International Tropical Timber Organization (ITTO), for example, have banned the export of raw logs to Japan (preferring instead to mill it domestically) and restrict exports so that they can institute sustainable forestry practices or, in the case of Thailand, prevent soil erosion and massive flooding. They have been joined by some core countries, for example, the Netherlands, which has proposed a ban on imported timber cut on a nonsustainable basis (Arden-Clarke 1990, 9). These would be illegal under new GATT provisions that prohibit import and export restrictions and illegal under new GATT rulings that prohibit nontariff barriers based on the method of manufacture.

By eliminating import and export restrictions, FTAs essentially guarantee continuing supplies of food, agricultural products, energy, and natural resources. Steady supplies make it easier for TNCs to plan production strategies and transfer the *costs* of shortages—either as a result of natural disaster or human consumption—to producers and workers and the environment in peripheral and core states.

Together, these FTA provisions are designed to improve the economic position of transnational corporations in the United States and other core countries, recapture market shares for large agricultural traders and agro-industries, improve the U.S. balance of trade, reduce the budget deficits of core states, facilitate the construction of flexible global commodity chains, and weaken groups that now receive financial assistance from states.

CRITIQUES OF FTAs

FTAs can be criticized on practical grounds, from the perspective of classical political economy, as well as from a world-system perspective. From a pragmatic perspective, even if FTAs result in substantial economic benefits that are widely shared, it is clear that consumers, producers, taxpayers, and the environment in both core and peripheral states will be injured in important ways. Cross-national coalitions of farmers, labor unions, small businesses, taxpayers, and consumer-rights and environmental-protection groups have therefore organized to defeat them (Davis 1992, 1).[32]

If the NAFTA and the Uruguay Round GATT proposals were not ratified—both have experienced political difficulties—many of their features would survive in other forms. The Super 301 section of the trade act can still be used as a crowbar to conclude bilateral agreements and voluntary measures that embody many of NAFTA and GATT's important provisions. The U.S.–Canada FTA would remain in effect. And GATT will remain in effect, even if the Uruguay Round amendments are not approved. Because GATT relies on case law and precedents, members are bringing suits that redefine the meaning of subsidies and nontariff trade barriers. In addition, other multilateral institutions, such as the World Bank and the United Nations, can be used to achieve many of the same objectives. When communist (China), capitalist (Taiwan), and former communist states (former Soviet republics) have applied for membership to GATT and other international organizations or have sought financial assistance from the World Bank or G-7 countries, core states have used the opportunity to press for agreements recognizing patent rights; eliminating state subsidies, regulations, and nontariff barriers; privatizing state assets; and adopting standards set by core countries.

The new FTAs can also be criticized from the perspective of classical political economists, for example, Adam Smith. This is important because FTA proponents argue that they promote free trade and that their economic benefits outweigh their deficiencies. From a Smithian perspective, however, the new FTAs actually violate three important principles of free trade.

As we have seen, FTAs tend to promote the interests of export traders and global agro-industrial enterprises at the expense of consumers, sometimes taxpayers, small businesses and farmers, and the environment (see McMichael 1993). It was this kind of development—the rise of what he called the "merchant-manufacturers"—that Smith attacked in *The Wealth of Nations*, arguing that "mercantilism" was anathema to expanded trade and economic competition.[33]

If one examines the industrial groups promoting FTAs and the trade representatives who negotiated the new provisions, there is considerable evidence that the interests of merchant-manufacturers were well represented, while the interests of other economic groups were neglected.

Because FTA provisions tend to promote the monopoly power of large TNCs, FTAs violate another fundamental Smithian principle: Free trade cannot coexist with monopoly. When he wrote *The Wealth of Nations*, Smith sided with the "interlopers" against the chartered monopolies that then dominated world trade, arguing that monopolies acted in ways to raise prices and stifle competition, which hurt small producers and consumers alike.[34]

Yet none of the new FTAs include any private antitrust provisions. If they were serious about free trade in the Smithian sense, one would think that U.S. negotiators would insist that features of U.S. antitrust law—the Sherman Act and its successors, which have been neglected of late—be written into the FTAs. Instead of insisting on antitrust provisions in the FTAs, the Bush administration actually "pledged to reform anti-trust laws to allow companies to enter joint production ventures [with Japanese firms]" (Sanger 1990, 1).[35] This is a profoundly un-Smithian approach.

In addition to these violations, the attempt to negotiate simultaneously both a global FTA (GATT) and regional FTAs (U.S.–Canada, NAFTA) is an oxymoron in Smithian terms because the latter establishes a kind of "protectionism" from the former.[36] The United States cannot grant most-favored-nation status to all GATT members and then offer a still more-favored nation status to Mexico and Canada without adversely affecting or discriminating against the former. What is more, this two-track strategy risks creating the kind of prewar trading blocks that the postwar GATT was supposed to prevent.[37]

Because they violate fundamental free trade principles, the new FTAs should be criticized as free trade counterfeits, a "pirated" version of the genuine Smithian article.

A third kind of critique can be made from a world-system perspective. There is little evidence that expanded world trade, which the new FTAs are supposed to facilitate, promotes economic development. As Arrighi (1991, 40) has demonstrated, the gap between core and peripheral states has widened since 1950. And global economic fortunes have diverged despite a tenfold expansion in the volume of world trade.

It is also important to recognize that the new FTAs represent a significant departure from the exploitative norms that characterized economic relations in the postwar period. They represent a new development, which I would characterize as "indifferent imperialism."

Indifferent imperialism is distinguished from previous forms of imperialism because compliance with relations of exploitation is achieved not by the direct administration policies of prewar colonialism or by the aid and intervention policies of neocolonialism in the postwar period, but by a system of rules and agreements that peripheral states themselves voluntarily adopt. They do so not because powerful states compel them by force, but because they will be excluded from global capitalist commodity chains if they do not. And, today, failure to participate means economic marginalization and ruin, not just for the poor, who have always been treated in this fashion by capitalism, but for the middle class and ruling strata as well, as is the case in much of Africa today.[38]

Indifferent imperialism is emerging because core countries are developing technologies that enable them to replace peripheral agricultural goods with industrial, biotechnological goods produced in the core, a process that Goodman (1991, 38–39) calls "substitutionism" and Friedmann (1991, 67) calls "import substitution." So, for example, the production of sugar by peripheral states is being replaced by high fructose corn sweeteners grown in the core (Friedmann 1991, 75–76), tropical oils by temperate oilseeds (Friedmann 1991, 77), livestock or fish protein by mycoprotein, and animal fats by soybeans (Goodman

1991, 47). These developments reduce core dependence on peripheral producers, a process that might be described as the disintermediation of primary producers.

Where peripheral products cannot be replaced, the core countries have encouraged competition among primary producers, either through loans or through different investment strategies. As Bunker and O'Hearn (1993) note, the Japanese have successfully diversified their sources of raw material and agricultural supplies by encouraging competition among producers around the world, which forces prices down and reduces the risk of shortage. Under these conditions, core countries have been able to disengage themselves from the kind of relations that led them to establish formal and informal systems of imperialism.

So peripheral states, like cities seeking professional sports franchises, compete desperately to obtain one of the limited number of opportunities to be exploited. In a sense, poor countries face a monopsony of core-country exploiters.

Indifferent imperialism is emerging because core countries have so greatly increased their technological advantages that they do not *need* to exploit the whole world, just some of it. And core states have discovered there are real economic and political benefits to be derived from reducing the administrative-military protection costs of achieving peripheral state compliance with relations of exploitation. For the first time, core countries can be indifferent as to who produces commodities, so long as someone does. And their monopsonistic position virtually guarantees that people around the world will compete ferociously for that right. Indifferent imperialism can be distinguished from the old colonial and neocolonial regulatory systems because it can be conducted without guilt or a sense of responsibility or obligation, what was once called the ''white man's burden.''

NOTES

1. Discussion of free trade agreements includes the Super 301 section of the Trade Acts of 1984 and 1988, the U.S.–Canada Free Trade Agreement, the North America Free Trade Agreement, the General Agreement on Tariffs and Trade, and the debt-crisis agreements negotiated by the World Bank with debtor states.

2. As Watkins (1992, 36) notes, ''By 1990, more than half of the 32 cases under Section 301 investigation involved developing countries, with eight of them, accused of failing to respect the intellectual property rights of American companies, being placed on a 'priority list.' '' See also Bradsher (1992).

3. As U.S. Trade Representative Carla Hills put it, ''There is no question about it. This round of GATT talks is a bold and ambitious undertaking. We want new rules governing investment, we want corporations to be able to make investments overseas without being required to take a local partner or export a given percentage of their output, to use local parts, or to meet any of a dozen other requirements'' (Tumulty and Havemann 1990, D1).

4. According to Weintraub (1991, 57), ''When it joined GATT in 1986, Mexico bound its tariff at 50 percent and said that it would gradually lower this rate. As of now, Mexico's average tariff is 9 percent, still more than that of the United States, but not

much higher than Canada's. It took both Canada and the United States some 40 years to reduce their tariffs by this magnitude. Mexico did it in less than five years.''

5. According to William Greider (1933, 35–36), "U.S. financial investment in Mexico, mostly stocks and bonds, went from zero in 1988 to $11 billion in 1992—a burgeoning new market for Wall Street. Cynics would note that U.S. investors will provide the capital that builds the new factories that will suck away U.S. jobs.'' (See also DePalma 1993, C1).

According to Rothstein (1933, 68), "General Motors is now Mexico's largest employer and half a million Mexicans now work in some 2,000 border plants (*maquiladoras*) that export to the United States.''

6. According to Princeton economist Gene Grossman, "You're removing a small amount of protection against a small country" (Nasar 1991, C2).

7. According to Romm and Lovins (1992/93, 47), "Oil imports alone have accounted for nearly three-fourths of the U.S. trade deficit since 1970, or $1 trillion transferred to OPEC nations.''

8. According to Friedmann (chapter 2), "In the United States, farm debt more than tripled in the 1970s, fueled by high prices and speculation in farm land.''

9. India increased grain production by nearly 50 percent, from 131.15 million metric tons in 1980 to 190.23 million metric tons in 1989 (Orr 1992, 65, 66). According to Watkins (1991, 40), "Rising self-sufficiency in key Asian markets, including India, Pakistan and Indonesia . . . further constrained import demand.''

10. According to Clairmonte (1991, 12), "U.S. agriculture has fallen on hard times, from 27 percent of world exports in the mid-1970s to 10 percent [in 1990].''

11. As Agricultural Secretary John Block said in 1986, "The push by some developing countries to become more self-sufficient in food may be reminiscent of a bygone era. These countries could save money by importing more food from the United States. . . . .The U.S. has used the World Bank to back up this policy, going so far as making the dismantling of farmer support programmes a condition for loans, as is the case of Morocco's support for their domestic cereal producers" (Ritchie 1987a, 5).

12. It fell to between $15 and $21 billion by 1988, largely because the dollar devaluation further cut U.S. prices on global markets. "Due to a combination of a weaker dollar and export subsidies, agricultural exports turned up again in quantity in 1987 and made further gains during 1988 in both quantity and value. But rising imports have kept the agricultural balance of trade lower than in the early 1980s,'' according to Porter and Bowers (1989, 19).

13. The Cairns group includes Australia, Canada, New Zealand, Fiji, Brazil, Uruguay, Argentina, Malaysia, Indonesia, Philippines, Thailand, Colombia, and Chile.

14. "There may be some manufacturers who are quietly hoping for the start of [rice] imports, but politically we cannot push for it,'' an official of the Sake brewers' association told the *New York Times* (Pollack 1992, 1).

In 1990, Japanese Prime Minister Toshiki Kaifu "grudgingly accepted, under strong pressure from President Bush, the idea of opening his country's rice market to foreign competition" (Nakagama 1990, C1).

15. According to some studies, a 20 percent reduction of agricultural subsidies is a kind of tipping point. A less than 20 percent reduction would not greatly hurt EC farmers, but a more than 20 percent cut would (Orr 1992, 111).

16. "In 1977 and 1982,'' notes Hipple (1990, 232), "nearly all of the U.S. trade

deficit could be linked to the intra firm transactions of foreign-based multinational companies.''

17. ''An important feature of the U.S.–Mexico industrial relationship is that more than one-half of their trade in manufactured goods takes place between affiliated companies. This is also true for U.S.–Canada industrial trade,'' notes Weintraub (1991, 59).

18. Indeed, Rene Osario, director of finance for Hewlet Packard's Latin American operations, told me, ''Duty reduction will free investment for other things.'' He did not say that it would enable HP to reduce consumer prices. (Interview, March 5, 1993.)

19. According to Shiva (1991, 11), ''A farmer purchasing [patented] seed would have the right to grow the seed but not the right to make seed.''

20. Under the 1961 Union for the Protection of New Varieties of Plants, which assigned the role of ''creator'' to plant breeders, ''not the country or farmers which provided the seed or discovered its use,'' core countries have appropriated much of the world's biological wealth. ''According to one estimate, over 55 percent of the world's collected germplasm is banked in the North—the U.S. alone holds 22 percent,'' even though ''Australia, Europe and North America combined meet less than six percent of their biotech needs for plant and animal species from their own resources'' (Agarwal and Narain 1992, 33).

21. ''One of the objectives for the Uruguay Round . . . is to 'minimize the adverse affect that sanitary and phyto-sanitary regulations and barriers can have on trade in agriculture, taking into account the relevant international agreements.' [But] these and other regulations and standards that protect public health and safety, the environment and national security are the only trade barriers that can be shown to benefit consumers'' (Bredahl and Forsythe 1989, 189).

22. ''In August 1991, a GATT panel ruled that nations may not restrict trade in products *on the basis of the process under which they are harvested or manufactured*,'' says Wallach (1991, 23) of the tuna decision.

23. U.S. Trade Representative Carla Hills complained that ''governments are looking for and finding alleged problems about [food] composition or how it was grown'' (Hills 1990).

24. ''The fight against regulation has become a fight against the states,'' argues Bruce Silverglade, director of legal affairs of the Washington-based Center for Science in the Public Interest. ''The business community has tamed the federal regulatory beast, so now the cry for further deregulation is synonymous with a cry for pre-emption'' (Moore 1990, 1758).

25. According to former California Governor Edmund G. (Jerry) Brown, Jr. (1992, 7), there is a ''fundamental conflict between intrusive international regulations and American democracy. Our constitutional system rests on democratic accountability with significant legal and regulatory differences recognized among states and localities. NAFTA . . . would curtail local preferences and thereby undermine the ability of diverse communities to control their destiny.''

26. ''The second goal of the food multinationals in the GATT talks is to make domestic supply management programs illegal or unworkable . . . [because] when farmers reduce their production in order to balance supplies with demand, they end up buying less fertilizer and pesticides,'' argues Ritchie (1990a, 27).

27. There is, of course, a lively struggle among these different producers for control over profit-making activity. In the United Kingdom, food retailers are using advances in

productivity and marketing to capture profits at the expense of food processors (Wrigley 1991).

28. "Low farm prices are not [going to help] consumers since the farmer's portion of the grocery bill has grown smaller and smaller with the expansion of the American food industry," argue Summers and Tufte (1988, 25–26).

29. In one proposal, trade-distorting domestic policies ("red light"), such as production subsidies, would be phased out in a ten-year period; mildly trade-distorting ("yellow light") policies, such as import subsidies, would be monitored and reduced; and minimal trade-distorting policies ("green light"), such as research and agricultural extension programs, disaster programs, and domestic food aid, would not be subject to GATT disciplines (McDonald 1990, 305).

30. Ritchie argues that "the U.S. is demanding a global phaseout of all agricultural import controls. This would be a disaster for food self-sufficiency [because] many food-deficit nations depend on import controls to insure that cheap, subsidized imports do not destroy their local farmers [and] import controls are absolutely necessary in over-producing countries, such as the United States and Europe, to allow effective supply management programs needed to prevent export dumping." Despite this, "U.S. officials have demanded that member nations: 'Agree to modify the GATT specifically in order to remove the ability of countries to restrict exports in times of critical food shortages' " (1988a, 3).

31. Officials also want to undo the damage done by the U.S. soy embargo of 1973, which demonstrated to Japan its vulnerability to foreign food suppliers (Friedmann chapter 2).

32. Six environmental groups, however, recently announced support for NAFTA: Environmental Defense Fund, National Audubon Society, National Wildlife Federation, Nature Conservancy, Natural Resources Defense Council, and World Wildlife Fund (Bradsher 1993a, C2). Former California Governor Jerry Brown and 1992 presidential candidate H. Ross Perot are the only two national political figures who have announced opposition to NAFTA.

33. "Liberty," according to Smith, "was sacrificed to monopoly, consumption was sacrificed to production, and the nation was sacrificed to merchant-manufacturers. Smith nurtured a tremendous antipathy to the merchant-manufacturers who, he claimed, falsely represented their interests as those of the nation. . . . [They] had succeeded in identifying the nation's interest with their own wealth. By capturing the state . . . they were able to shape the market and regulate trade to their own ends—ends contrary to those of consumers, the people, the nation" (Schaeffer 1981, 83).

34. For example, speaking of the English East India Company, Smith argued, "By a perpetual monopoly, all the other subjects of the state are taxed very absurdly in two different ways: first, by the high price of goods, which, in the case of free trade, they could buy much cheaper; and secondly, by their total exclusion from a branch of business, which it might be both convenient and profitable for many of them to carry on" Smith 1976, 278).

35. At the time, Vice President Dan Quayle argued, "To make America more competitive, we are . . . going to have to reexamine our anti-trust laws, many of which are anachronistic in this age of global competition" (Davidow 1991, 38).

36. As Smith argues in "Of Treaties of Commerce," "When a nation binds itself to a treaty . . . to exempt the goods of one country from duties to which it subjects those of all others [NAFTA, for example] . . . [t]hose merchants and manufacturers enjoy a sort

of monopoly in the country which is so indulgent of them, . . . [and sell] their goods for a better price than if exposed to the free competition of all nations. Such treaties, however, though they may be advantageous to the merchants and manufacturers of the favoured, are necessarily disadvantageous to those of the favoring country'' (1976, II: 53).

37. "The lax enforcement of Article XXIV [which prohibits bilateral and regional trade agreements that raise barriers to third-country trade] has 'set a dangerous precedent for further special deals, fragmentation of the trading system, and damage to the trade interests of non-participants,' " argues Leutwiler in Schott (1991, 3).

38. According to Attali (1991, 73), "Europe's periphery, Africa, is a lost continent. It is one of the last places on earth in which famine persists. The terrible facts of having fallen into an economic black hole speak for themselves: since 1970, Africa's share of the world markets have been reduced by half; its debt has been multiplied by twenty and now equals its total gross product; and income per capita in sub-Saharan Africa has fallen by one-quarter since 1987.''

REFERENCES

Agarwal, Nail, and Sunita Narain. 1992. "A Royalty for Every Potato." *Earth Island Journal*, Winter: 33.

Arden-Clark, Charles. 1990. *Conservation and Sustainable Management of Tropical Forests: The Role of ITTO and GATT*. Gland, Switzerland: World Conservation Center.

Arrighi, Giovanni. 1991. "World Income Inequalities and the Future of Socialism." *New Left Review* 189: 39–65.

Attali, Jacques. 1991. *Millennium: Winners and Losers in the Coming World Order*. New York: Times Books.

Bhagwati, Jagdish. 1989. "Super 301's Big Bite Flouts the Rules." *New York Times*, June 4, C2.

Bradsher, Keith. 1991. "Bush Tells Mexican Leader He Wants Trade Pact Soon." *New York Times*, December 15, A10.

———. 1992. "U.S. Adds Seven Countries to Trade Barrier List." *New York Times*, March 31, C2.

———. 1993a. "Free Trade Pact Wins Wide Range of Support." *New York Times*, April 30, C2.

———. 1993b. "A Look at the North American Pact's Added Costs." *New York Times*, July 14, C1.

Bredahl, Maury E., and Kenneth W. Forsythe. 1989. "Harmonizing Phyto-sanitary and Sanitary Regulations." *World Economy* 12, 2: 189–206.

Brown, Edmund G., Jr. 1992. "Race to the Bottom." *San Jose Mercury News*, September 15, 7B.

Bunker, Stephen G., and Denis O'Hearn. 1993. "Strategies of Economic Ascendants for Access to Raw Materials: A Comparison of the United States and Japan. In *Pacific Asia and the Future of the World-System*, edited by Ravi Palat, 83–102. Westport, Conn.: Greenwood Press.

Chomsky, Noam. 1993. "The Masters of Mankind." *The Nation*, March 29, 412–416.

Christensen, Eric. 1991. *GATT Nets an Environmental Disaster: A Legal Analysis and Critique of the GATT Panel Ruling on Imports of Yellow Fin Tuna into the United States*. Washington, D.C.: Community Nutrition Institute.

Christensen, Eric, and Samantha Geffin. 1991–92. "GATT Sets Its Net on Environmental Regulation: The GATT Panel Ruling on Mexico Yellow Fin Tuna Imports and the Need for Reform of the International Trading System." *Inter-American Law Review* 23, 2: 570–612.

Clairmonte, Frederick. 1991. "The Debacle of the Uruguay Round: An Autopsy." *Third World Economy*, January 16–31.

Compa, Lance. 1992. "Pre-emption of State Law." In *US Citizens' Analysis of the NAFTA*. Washington, D.C.: The Development GAP.

Davidow, Joel. 1991. "The Relationship between Anti-trust and Trade Laws in the United States." *World Economy* 14 (1): 37–51.

Davis, Bob. 1992. "Fighting NAFTA: Free Trade Pact Spurs a Diverse Coalition of Grass-roots Foes." *Wall Street Journal*, December 23, 1.

Deardorff, Alan V. 1990. "Should Patent Protection Be Extended to All Developing Countries?" *World Economy* 13, 4: 497–507.

DeLind, Laura B. 1992. "Cheap Food: A Case of Mind over Matter." Paper presented for the Conference on Diversity in Food, Agriculture, Nutrition and Development. East Lansing, Michigan, June 4–7.

DePalma, Anthony. 1993a. "Fortunes Are Cast in Mexican Stocks." *New York Times*, April 11, D1.

Farnsworth, Clyde H. 1992. "New Trade War Target May Be Dogs." *New York Times*, December 2, D1.

Friedmann, Harriet. 1991. "Changes in the International Division of Labor: Agri-food Complexes and Export Agriculture." In *Towards a New Political Economy of Agriculture*, edited by William H. Friedland, Lawrence Busch, Frederick H. Buttel, and Alan P. Rudy, 65–93. Boulder, Colo.: Westview Press.

Goodman, David. 1991. "Some Recent Tendencies in the Industrial Reorganization of the Agri-Food System. In *Towards a New Political Economy of Agriculture*, edited by William H. Friedland, Lawrence Busch, Frederick H. Buttel, and Alan P. Rudy, 37–64. Boulder, Colo.: Westview Press.

Greenpeace. 1992. *Green Fields, Grey Future: EC Agricultural Policy at the Crossroads.* Amsterdam: Greenpeace International.

Greider, William. 1993. "Clinton's Choice: Jobs or Mexico." *Rolling Stone*, May 13, 35–36.

Grimmet, Jeanne J. 1991. *Environmental Regulation and the GATT.* Washington, D.C.: Congressional Research Service.

Hills, Carla. 1990. "New Threats to Agriculture Exports Seen Under Guise of Food Safety." *Inside U.S. Trade*, March 16, 5–8.

Hipple, F. Steb. 1990. "Multinational Companies and the Growth of the U.S. Trade Deficit." *International Trade Journal* 5, 2: 217–34.

Hufbauer, Gary. 1989. *The Free Trade Debate: Reports of the Twentieth Century Fund Task Force on the Future of American Trade Policy.* New York: Priority Press Publications.

Jerome, Robert W. 1992. *World Trade at the Crossroads: The Uruguay Round, GATT and Beyond.* New York: University Press of America.

Johnstone, Diana. 1992. "Last Ditch or Frontline?" *In These Times*, December 28: 24–25.

McDonald, Bradley J. 1990. "Agricultural Negotiations in the Uruguay Round." *World Economy* 13, 3: 299–326.

McMichael, Philip. 1993. "World Food System Restructuring under a GATT Regime." *Political Geography* 12, 3: 198–214.

Moore, W. John. 1990. "Stopping the States." *National Journal*, July 21, 1758–62.

Morici, Peter. 1993. "Grasping the Benefits of NAFTA." *Current History* 92, 571: 49–54.

Nakagama, Sam. 1990. "In Japan, Farm Supports Prop up More Than Farms." *New York Times*, July 13, C1.

Nasar, Sylvia. 1992. "Trade Pact Fears Seem Overstated." *New York Times*, August 6, C2.

Office of Management and Budget. 1993. "U.S. Budget Trends." *San Francisco Chronicle*, April 9, A4.

Office of the U.S. Trade Representative. 1986. "American Concerns." In *The Free Trade Papers*, edited by Duncan Cameron, 48–62. Toronto, Ontario: James Lorimer and Company.

Orr, Bill. 1992. *The Global Economy in the '90s: A Users' Guide.* New York: New York University Press.

Pollack, Andrew. 1992. "A Rice-short Japan Reconsiders the Tradition of Banning Imports." *New York Times*, August 17.

Porter, Jane M., and Douglas E. Bowers. 1989. *A Short History of U.S. Agricultural Trade Negotiations.* Staff Report no. AGES 89-23. Washington, D.C.: Agricultural and Rural Economy Division, Economic Research Service, U.S. Department of Agriculture.

Raghavan, Chakravarthi. 1990. *Recolonization: GATT, the Uruguay Round and the Third World.* Penang, Malaysia: Third World Network.

Ritchie, Mark. 1987a. "Alternatives to Agricultural Trade War." *CAP Briefing*, Nos. 4–5. London: Catholic Institute for International Relations.

———. 1987b. *Crisis by Design: A Brief Review of U.S. Farm Policy.* Minneapolis: League of Rural Voters Education Project.

———. 1988a. *The Decoupled Approach to Agriculture.* Minneapolis: Institute for Agriculture and Trade Policy.

———. 1988b. *Impact of GATT on Food Self-Reliance and World Hunger.* Minneapolis: Institute for Agriculture and Trade Policy.

———. 1989. *The Environmental Implications of the GATT Negotiations.* London: Rongead.

———. 1990a. "GATT, Agriculture and the Environment: The US Double Zero Plan." *The Ecologist* 20, 6: 214–20.

———. 1990b. *Free Trade vs Sustainable Agriculture.* Minneapolis: Institute for Agriculture and Trade Policy.

———. 1990c. "Trading Away the Family Farm: GATT and Agriculture." *Multinational Monitor*, November, 26–28.

Romm, Joseph J., and Amory B. Lovins. 1992/93. "Fueling a Competitive Economy." *Foreign Affairs*, Winter, 46–62.

Rongead. 1990. "Of Minds and Markets: Intellectual Property Rights and the Poor." *GATT Briefing*, no. 2, July, European Network on Agriculture and Development.

Rosset, Peter M. 1991. *Non-traditional Export Agriculture in Central America: Impact on Peasant Farmers.* Paper presented at the Workshop on the Globalization of the Fresh Fruit and Vegetable Industry, University of California, Santa Cruz, December 6–9.

Rothstein, Richard. 1993. "Continental Drift: NAFTA and Its Aftershocks." *American Prospect* 12: 68–84.

Sanger, David E. 1990. "U.S. and Japan Set Accord to Rectify Trade Imbalances." *New York Times*, June 29, A1.

Schaeffer, Robert. 1981. "The Entelechies of Mercantilism." *Scandinavian Economic History Review* 29, 2: 81–96.

———. 1993. "Democratic Devolutions: East Asian Democratization in Comparative Perspective." In *Pacific Asia and the Future of the World-System*, edited by Ravi Palat, 169–80. Westport, Conn.: Greenwood Press.

———. Forthcoming. "Standardization, GATT and the Fresh Food System." *International Journal of Sociology of Agriculture and Food*.

Schott, Jeffrey J. 1991. "Trading Blocs and the World Trading System." *World Economy* 14, 1: 1–17.

Shiva, Vanda. 1991. "The Crisis of Diversity." *Third World Resurgence* 13: 10.

Shrybman, Steven. 1990. "The Environmental Costs of Free Trade." *Multinational Monitor*, March, 20–22.

Skidmore, David. 1992. "'92's Record U.S. Deficit." *San Francisco Chronicle-Examiner*, October 29, A1.

Smith, Adam. 1976. *An inquiry into the Nature and Causes of the Wealth of Nations*. Vols. I and II. Chicago: University of Chicago Press.

Soden, Katie. 1988. *US Farm Subsidies*. Snowmass, Colo.: Rocky Mountain Institute Agriculture Program.

Stanford, Lois. 1991. *Mexico's Fresh Fruit and Vegetable Export System: Recent Developments and Their Impact on Local Economies*. Paper presented at the Workshop on the Globalization of the Fresh Fruit and Vegetable Industry, University of California, Santa Cruz, December 6–9.

Summers, Mary, and Edward Tufte. 1988. *The Crisis of American Agriculture: Minding the Public Business*. Minneapolis: Institute for Agriculture and Trade Policy.

Tarditi, Secondo, Kenneth J. Thomson, Pierpaolo Pierani and Elisabetta Croci-Angelini. *Agricultural Trade Liberalization and the European Community*. Oxford, England: Clarendon Press.

Tester, Frank James. 1988. "Free Trading the Environment." In *The Free Trade Deal*, edited by Duncan Cameron, 197–214. Toronto, Ontario: James Lorimer and Company.

Tumulty, Karen and Joel Havemann. 1990. "Third World Nations Complain That Trade Talks Exclude Them." *Los Angeles Times*, December 6, D1.

Wallach, Lori. 1991. *Consumer and Environmental Analysis: Uruguay Round of the GATT*. Washington, D.C.: Public Citizen.

Warnock, John W. 1988. "Agriculture and the Food Industry." In *The Free Trade Deal*, edited by Duncan Cameron, 166–75. Toronto, Ontario: James Lorimer and Company.

Watkins, Kevin. 1989. "Agriculture and Farm Trade in the GATT." *CAP Briefing*, 20. London: Catholic Institute for International Relations.

———. 1990. "Changing the Rules: The GATT Farm Trade Reform and World Food Security." *GATT Briefing on Agriculture, Food Security and the Uruguay Round*. London: European Network on Agricultural Development, no. 4.

———. 1991. "Agriculture and Food Security in the GATT Uruguay Round." *Review of African Political Economy* 50: 38–50.

————. 1992. "GATT and the Third World." *Race and Class* 34, 1: 23–40.
Weintraub, Sidney. 1991. "Free Trade in North America: Has Its Time Come?" *World Economy* 14, 1: 57–66.
————. 1993. "The Economy on the Eve of Free Trade." *Current History* 92, 571: 67–72.
Wrigley, Neil. 1991. "Retail Power: The Context of the Fresh Fruit and Vegetable System in the United Kingdom." Paper prepared for the Workshop on the Globalization of the Fresh Fruit and Vegetable System, University of California, Santa Cruz, December 6–9.

Selected Bibliography

Barkin, D., R. Batt, and B. DeWalt. 1990. *Feed Crops vs. Food Crops: The Global Substitution of Grains in Production.* Boulder, Colo.: Lynne Reiner.

Beckford, G. L. 1972. *Persistent Poverty: Underdevelopment in Plantation Economies in the Third World.* New York: Oxford University Press.

Bernstein, H., B. Crow, M. Mackintosh, and C. Martin, eds. 1990. *The Food Question: Profits versus People?* New York: Monthly Review Press.

Bonanno, A., L. Busch, W. Friedland, L. Gouveia, and E. Mingione, eds. *From Columbus to Conagra: The Globalization of Agriculture and Food.* Lawrence: Kansas University Press.

Bunker, S. G. 1987. *Peasants against the State: The Politics of Market Control in Bugisu, Uganda, 1900–1983.* Urbana: University of Illinois Press.

Burbach, R., and P. Flynn. 1980. *Agribusiness in the Americas.* New York: Monthly Review Press.

Cooper, F., et al. 1993. *Confronting Historical Paradigms: Peasants, Labor and the Capitalist World System in Africa and Latin America.* Madison: University of Wisconsin Press.

de Janvry, A. 1981. *The Agrarian Question and Reformism in Latin America.* Baltimore: Johns Hopkins University Press.

Denoon, D. 1983. *Settler Capitalism: The Dynamics of Dependent Development in the Southern Hemisphere.* Oxford, England: Clarendon Press.

Feder, E. 1978. *Strawberry Imperialism: An Inquiry into the Mechanisms of Dependency in Mexican Agriculture.* Mexico City: Editorial Campesina.

Friedland, W., L. Busch, F. H. Buttel, and A. Rudy, eds. *Towards a New Political Economy of Agriculture.* Boulder, Colo.: Westview Press.

Friedmann, H. 1981. "The Political Economy of Food: The Rise and Fall of the Postwar International Food Order." *American Journal of Sociology* 88: 248–86.

———. 1987. "Family Farms and International Food Regimes." In *Peasants and Peasant Societies,* edited by T. Shanin, 247–58. Oxford, England: Basil Blackwell.

————. 1992. "Distance and Durability: Shaky Foundations for the World Food Economy." *Third World Quarterly* 13, 2: 371–83.

————. 1993. "The Political Economy of Food: A Global Crisis." *New Left Review* 197: 29–58.

Friedmann, H., and P. McMichael. 1989. "Agriculture and the State System: The Rise and Decline of National Agricultures, 1870 to the Present." *Sociologia Ruralis* 29, 2: 93–117.

Goodman, D., and M. Redclift. 1981. *From Peasant to Proletarian: Capitalist Development and Agrarian Transitions*. Oxford, England: Basil Blackwell.

————, eds. 1990. *The International Farm Crisis*. New York: St. Martin's Press.

Goodman, D., B. Sorj, and J. Wilkinson. 1987. *From Farming to Biotechnology: A Theory of Agro-Industrial Development*. Oxford, England: Basil Blackwell.

Hathaway, D. E. 1987. *Agriculture and the GATT: Rewriting the Rules*. Washington, D.C.: Institute for International Economics.

Kloppenburg, J., Jr. 1988. *First the Seed: The Political Economy of Plant Biotechnology*. Cambridge, England: Cambridge University Press.

Le Heron, R. 1993. *Globalized Agriculture*. Oxford, England: Pergamon Press.

Le Roy Ladurie, E., and J. Goy. 1982. *Agricultural Fluctuations*. Cambridge, England: Cambridge University Press.

Llambi, L. 1990. "Transitions to and within Capitalism: Agrarian Transitions in Latin America." *Sociologia Ruralis* 30, 2: 174–96.

Mackintosh, M. 1989. *Gender, Class and Rural Transition: Agribusiness and the Food Crisis in Senegal*. London: Zed Books.

McMichael, P. 1984. *Settlers and the Agrarian Question: Foundations of Capitalism in Colonial Australia*. Cambridge, England: Cambridge University Press.

————. 1992. "Tensions between National and International Control of the World Food Order: Contours of a New Food Regime." *Sociological Perspectives* 35, 2: 343–65.

————. 1993. "World Food System Restructuring under a GATT Regime." *Political Geography* 12, 3: 198–215.

————, ed. 1994. *The Global Restructuring of Agro-Food Systems*. Ithaca, N.Y.: Cornell University Press.

McMichael, P., and D. Myhre. 1991. "Global Regulation vs. the Nation-State: Agro-Food Systems and the New Politics of Capital." *Capital & Class* 43: 83–105.

Malenbaum, W. 1953. *The World Wheat Economy, 1885–1913*. London: George Allen & Unwin.

Mintz, S. 1974. *Caribbean Transformations*. Chicago: Aldine.

————. 1985. *Sweetness and Power: The Place of Sugar in Modern History*. New York: Penguin.

Moore, B. 1967. *Social Origins of Dictatorship and Democracy: Lord and Peasant in the Making of the Modern World*. Boston: Beacon.

Morgan, D. 1980. *Merchants of Grain*. Harmondsworth, England: Penguin.

Paige, J. 1975. *Agrarian Revolution: Social Movements and Export Agriculture in the Underdeveloped World*. New York: Free Press.

Polanyi, K. 1957. *The Great Transformation: The Political and Economic Origins of Our Time*. Boston: Beacon.

Raghavan, C. 1990. *Recolonization: GATT, the Uruguay Round and the Third World*. Penang, Malaysia: Third World Network.

Raikes, P. 1988. *Modernising Hunger: Famine, Food Surplus & Farm Policy in the EEC and Africa*. London: Catholic Institute for International Affairs.

Raynolds, L., D. Myhre, P. McMichael, V. Carro-Figueroa, and F. H. Buttel. 1993. ''The 'New' Internationalization of Agriculture: Critique and Reformulation.'' *World Development* 21, 7: 1101–21.

Roseberry, W. 1983. *Coffee and Capitalism in the Venezuelan Andes*. Austin: University of Texas Press.

Sanderson, S., ed. 1985. *The Americas in the New International Division of Labor*. New York: Holmes and Meier.

———. 1986. *The Transformation of Mexican Agriculture: International Structure and the Politics of Rural Change*. Princeton, N.J.: Princeton University Press.

Skocpol, T. 1979. *States and Social Revolutions: A Comparative Analysis of France, Russia and China*. Cambridge, England: Cambridge University Press.

Stoler, A. L. 1985. *Capitalism and Confrontation in Sumatra's Plantation Belt, 1870–1979*. New Haven, Conn.: Yale University Press.

Tomich, D. 1990. *Slavery in the Circuit of Sugar: Martinique and the World Economy 1830–1848*. Baltimore: Johns Hopkins University Press.

Trouillot, M-R. 1988. *Peasants and Capital: Dominica in the World Economy*. Baltimore: Johns Hopkins University Press.

Wallerstein, I. 1974. *The Modern World-System*. Vol. 1, *Capitalist Agriculture and the Origins of the European World-Economy in the Sixteenth Century*. New York: Academic Press.

Wallerstein, M. B. 1980. *Food for War—Food for Peace*. Cambridge, Mass.: MIT Press.

Zamosc, L. 1986. *The Agrarian Question and the Peasant Movement in Colombia*. Cambridge, England: Cambridge University Press.

Index

About the Contributors

LANFRANCO BLANCHETTI-REVELLI is a doctoral student in Anthropology at Johns Hopkins University. He has done extensive fieldwork in Southern Palawan, Philippines, first among the Molbog, and more recently among Tausug and Samal migrants involved in seaweed cultivation. He is currently finishing a dissertation on the Political Economy of Seaweed.

JANE L. COLLINS is Associate Professor of Sociology and Women's Studies at the University of Wisconsin, Madison. She is the author of *Unseasonal Migrations: The Effects of Rural Labor Scarcity in Peru*, as well as many articles on labor, agriculture, and women's issues. She is co-editor of several volumes including *Work Without Wages* (SUNY Press) (with Martha Gimenez). She is currently completing a research project on gender and labor issues in export fruit production in northeastern Brazil.

J. TADLOCK COWAN is Assistant Professor of Sociology at Bates College in Lewiston, Maine, where he teaches courses in the sociology of science and technology, environment and development, and political sociology. His principal research interests are in regional development, the social impacts of new technologies, and economic restructuring.

HARRIET FRIEDMANN is Professor of Sociology at the University of Toronto. Currently her research focuses on cultural and material aspects of diets in relation to sustainable food economies, and on the place of food within the Cold War and post–Cold War shifts in international power and property relations. Her most recent publications are in *New Left Review, Food for the Future*, and the *Bulletin of the Institute of Development Studies*.

WALTER GOLDFRANK teaches sociology at the University of California, Santa Cruz, where he is also Provost of College VIII. He currently coordinates a project studying the social and ecological consequences of the transition to export fruit cultivation in the Aconcgua Valley.

RAMON GROSFOGUEL is a Postdoctoral Fellow at the Fernand Braudel Center (SUNY-Binghamton). He is editing a book about Puerto Rico entitled *Beyond Nationalism and Colonialism: Rethinking the Puerto Rican Political Imaginary.*

CRAIG K. HARRIS is Associate Professor in the Department of Sociology at Michigan State University, specializing in environmental sociology and rural sociology. His research on fisheries management and development has included the North American Great Lakes, the Senegalese coastal fishery, ocean fisheries of Taiwan, and the Great Lakes of East Africa. His research on global commodity chains has included fruit crops and pest management technologies.

RESAT KASABA is Associate Professor of International Studies and Adjunct Professor of Sociology at the University of Washington. He is the author of *The Ottoman Empire and the World Economy, The Nineteenth Century.* He has also edited *The Cities in the World-System.*

MIGUEL E. KORZENIEWICZ teaches sociology at the University of New Mexico at Albuquerque. He is the co-editor (with Gary Gereffi) of *Commodity Chains and Global Capitalism* (Greenwood Press, 1993). Other recent publications have appeared in *Sociological Perspectives* and several edited books.

ROBERTO P. KORZENIEWICZ is Assistant Professor of Sociology at the University of Maryland at College Park. His most recent publications have appeared in the *Hispanic American Historical Review,* the *Latin American Research Review,* and *Sociological Forum,* as well as in several edited books.

PHILIP McMICHAEL is Associate Professor of Rural and Development Sociology. He is the author of *Settlers and the Agrarian Question* and editor of *The Global Restructuring of Agro-Food Systems.* His research interests include historical sociology of the world economy, global regulation, and current agro-food system restructuring in the Pacific Rim.

SIDNEY MINTZ is the Wm. L. Straus, Jr. Professor of Anthropology, Johns Hopkins University, and a specialist in the Caribbean region. His books include *Worker in the Cane, Caribbean Transformations, Sweetness and Power,* and most recently (with Richard Price) *The Birth of African American Culture.*

RAVI ARVIND PALAT is Lecturer in Sociology at the University of Auckland, New Zealand and editor of *Pacific-Asia and the Future of the World-System.*

ROBERT SCHAEFFER teaches global sociology at San Jose State University in California. He is the author of *Warpaths: The Politics of Partition* and editor of *War in the World-System.* He is currently writing a book on contemporary global problems.

MIKE SKLADANY is currently a doctoral student in the Department of Sociology at Michigan State University, where he is conducting his dissertation research on the social construction of science and technology in aquaculture. Prior to attending graduate school, he worked in rural coastal Thailand with small-scale fishers and aquaculturists.

FARUK TABAK is a Research Associate at the Fernand Braudel Center, Binghamton University. He has edited (with Caglar Keyder) *Landholding and Commercial Agriculture in the Middle East.*

JOHN M. TALBOT is a Ph.D. candidate in sociology at UC-Berkeley, currently writing a dissertation on the Political Economy of the World Coffee Market.

SUSAN J. THOMPSON is a Visiting Assistant Professor of Sociology at Bates College in Lewiston, Maine. Her primary areas of research include regional food systems, the social organization of water resource development, and development policies in south and southeast Asia.

FRANCES M. UFKES is Assistant Professor in Geography at Dartmouth University. Her research interests include U.S. meat industry restructuring and the transformation of agricultural and food systems in Southeast Asia. She edited a special issue of *Political Geography* on the globalization of agriculture.

Studies in the Political Economy of the World-System
(Formerly published as Political Economy of the World-System Annuals)

Numbers 1–6 published by Sage Publications.

ISBN 0-313-29399-6

90000>

EAN

9 780313 293993

HARDCOVER BAR CODE